Water and Food Security in Central Asia

NATO Science for Peace and Security Series

This Series presents the results of scientific meetings supported under the NATO Programme: Science for Peace and Security (SPS).

The NATO SPS Programme supports meetings in the following Key Priority areas: (1) Defence Against Terrorism; (2) Countering other Threats to Security and (3) NATO, Partner and Mediterranean Dialogue Country Priorities. The types of meeting supported are generally "Advanced Study Institutes" and "Advanced Research Workshops". The NATO SPS Series collects together the results of these meetings. The meetings are coorganized by scientists from NATO countries and scientists from NATO's "Partner" or "Mediterranean Dialogue" countries. The observations and recommendations made at the meetings, as well as the contents of the volumes in the Series, reflect those of participants and contributors only; they should not necessarily be regarded as reflecting NATO views or policy.

Advanced Study Institutes (ASI) are high-level tutorial courses intended to convey the latest developments in a subject to an advanced-level audience

Advanced Research Workshops (ARW) are expert meetings where an intense but informal exchange of views at the frontiers of a subject aims at identifying directions for future action

Following a transformation of the programme in 2006 the Series has been re-named and re-organised. Recent volumes on topics not related to security, which result from meetings supported under the programme earlier, may be found in the NATO Science Series.

The Series is published by IOS Press, Amsterdam, and Springer, Dordrecht, in conjunction with the NATO Public Diplomacy Division.

Sub-Series

A.	Chemistry and Biology	Springer
B.	Physics and Biophysics	Springer
C.	Environmental Security	Springer
D.	Information and Communication Security	IOS Press
E.	Human and Societal Dynamics	IOS Press

http://www.nato.int/science
http://www.springer.com
http://www.iospress.nl

Series C: Environmental Security

Water and Food Security in Central Asia

edited by

Chandra A. Madramootoo

Faculty of Agricultural and Environmental Sciences
McGill University, Montreal, Canada

and

Viktor A. Dukhovny

Scientific Information Center of the Interstate Commission for Water Coordination, Tashkent, Republic of Uzbekistan

technical editors

Robert S. Baker

Brace Centre for Water Resources Management
McGill University, Montreal, Canada

and

I. Helen Fyles

Brace Centre for Water Resources Management
McGill University, Montreal, Canada

Springer

Published in cooperation with NATO Public Diplomacy Division

Proceedings of the NATO Advanced Research Workshop on
Socio-Economic Stability and Water Productivity:
Food and Water Security in Central Asia
Tashkent, Uzbekistan
18-20 March 2008

ISBN 978-94-007-0112-0 (PB)
ISBN 978-90-481-9973-0 (HB)
ISBN 978-90-481-9974-7 (e-book)

Published by Springer,
P.O. Box 17, 3300 AA Dordrecht, The Netherlands.

www.springer.com

Printed on acid-free paper

Preface

The purpose of this publication is to provide insight into the growing challenges of water and food security in Central Asia. This region is located in semiarid and arid vegetation zones, and large parts of its economy depend on water for irrigation and energy. As requirements for food production grow due to a rapidly increasing population and the industrial sector expands, the pressure on water resources will intensify. In addition, the likelihood of droughts due to climate change will put further demands on already limited water resources. Each chapter of this book provides necessary reading for those who seek a better understanding of the challenges surrounding water and food security in Central Asia. The authors, including researchers and scientists from the five Central Asian countries of Kazakhstan, Kyrgyzstan, Tajikistan, Turkmenistan and Uzbekistan, as well as international experts provide perspectives and recommendations based on decades of research and first hand experience.

Global, country and regional perspectives and strategies are presented on topics including: integrated water resources management, modernization of irrigation and drainage systems, legal framework and economic mechanisms for water resources management, regional and national water development strategies, climate change, water saving technologies and incentives, training of engineers, extension services for farmers, and water user associations. The linkage between irrigated agriculture, food self sufficiency, environmental security, and socio-economic stability is a general theme throughout this book.

The editors extend their deep appreciation to all the authors; this publication would not have been possible without their input and hard work. We are most grateful to Springer for affording the opportunity to publish this volume. We thank the editorial and publishing staff at Springer for their assistance, organization and efficiency, in particular, we would like to express our gratitude to Annelies Kersbergen and Wil Bruins for their help. We would also like to express our sincere thanks to the North Atlantic Treaty Organization for their support of the Advanced Research Workshop entitled "Socio-Economic Stability and Water Productivity: Implications of Food and Water Security in the Central Asian Region" which served as the original thrust for this publication.

Contents

Contributors

Myratgheldy Akmammedov
Minister of Water Economy of Turkmenistan, Deputy Chairman
of the Cabinet of Ministers of Turkmenistan 142 Hero of Turkmenistan
A. Niyazov Str., Ashgabat 744000, Turkmenistan

Atalyev Kakadurdy
Ministry of Water Economy of Turkmenistan 142 Hero of Turkmenistan,
A. Niyazov Str., Ashgabat 744000, Turkmenistan

Agajan Geldiyevich Babaev
Interstate Commission for Sustainable Development of Turkmenistan
142 Hero of Turkmenistan, A. Niyazov Str.,, Ashgabat 744000, Turkmenistan

Robert Baker
Brace Centre for Water Resources Management, McGill University,
Montreal, Canada

Sergei Yevstaphievich Bednaruk
Ministry of Natural Resources of the Russian Federation, Center of Russian
Waterworks Inventory and State Water Cadastre 2/2-a, Pryanishnikov Str., Moscow
127550, Russian Federation

Ismail Begimov
Scientific Information Centre of Interstate Commission for Water Coordination
B. 11, Karasu-4, Tashkent 700187, Republic of Uzbekistan

Iskander Ferdinandovich Beglov
Scientific Information Center of Interstate Commission for Water Coordination
B. 11, Karasu-4, Tashkent, 700187, Republic of Uzbekistan

Victor A. Dukhovny
Scientific Information Center of Interstate Commission for Water Coordination (SIC ICWC), B.11, Karasu-4, Tashkent 700187, Republic of Uzbekistan

Kurbanbaev Erejep
Karakalpak Branch, Central Asian Research Institute of Irrigation (SANIIRI), 12 Dustlik Str., Nukus 742015, Republic of Uzbekistan

Helen Fyles
Brace Centre for Water Resources Management, McGill University, Montreal, Canada

Nazar Ibragim
Uzbek Cotton Research Institute Tashkent

Timur Kamalovich Kamalov
State Committee for Safety of Large-scale Hydro-structures 7 Navoiy Str., Tashkent 700011, Republic of Uzbekistan

Makhmud Khamidov
Basin Water Organization Syrdarya B.11, Karasu-4, Tashkent 700187, Republic of Uzbekistan

Shavkat R. Khamraev
Ministry of Agriculture and Water Resources of Uzbekistan 4 Navoiy Str., Tashkent, 100004, Republic of Uzbekistan

Yuldash Khudaybergnov
Basin Water Organization Amudarya 63 Al-Zamakhshary Str., 140000, Urgench Republic of Uzbekistan

Kienzler Kirsten
International Center for Agricultural Research in the Dry Areas (ICARDA), Regional office for Central Asia and the Caucasus (CAC) P.O. Box 4564, 100000, Tashkent Uzbekistan

Barataly Turanovich Koshmatov
Ministry of Agriculture, Water Resources and Processing Industry of the Kyrgyz Republic, Water Resources Department, 4a Toktonaliev Str., Bishkek 720055, Kyrgyz Republic

John P.A. Lamers
German-Uzbek Landscape Restructuring Project, Center for Development Research (ZEF) Walter-Flex-Str. 3, Bonn 53113, Germany

Chandra Madramootoo
Faculty of Agricultural and Environmental Sciences, McGill University, Macdonald Campus, Ste-Anne-de-Bellevue, Quebec, Canada

Sergei Alexeyevich Nerozin
SIC ICWC, Regional expert, B.11, Karasu-4, Tashkent 700187, Republic of Uzbekistan

Anatoliy Dmitriyevich Ryabtsev
Committee for Water Resources, Ministry of Agriculture, 9 Avenue Pobeda, Astana, 010000, Republic of Kazakhstan

Aly Shady
Natural Resources Division, Canadian International Development Agency, 200 Promenade du Portage, Hull, Quebec, Canada

Vadim Sokolov
Scientific Information Center of Interstate Commission for Water Coordination (SIC ICWC), B.11, Karasu-4, Tashkent 700187, Republic of Uzbekistan

Anatoly Sorokin
Scientific Information Center of Interstate Commission for Water Coordination (SIC ICWC), Tashkent

Stulina Galina
Scientific Information Center of the Interstate Commission for Water Coordination B.11, Karasu-4, Tashkent 700187, Rebublic of Uzbekistan

Pulatkhon Umarov
Scientific Information Center of Interstate Commission for Water Coordination (SIC ICWC), Training Centre of ICWC, B.11, Karasu-4, Tashkent 700187, Republic of Uzbekistan

Said Yokubzod
Minister of Land Reclamation and Water Resources of the Republic of Tajikistan 5/1 Shansi Str., Dushanbe 734064, Republic of Tajikistan

Karimova Olga Yushevna
Karakalpak Branch, Central Asian Research Institute of Irrigation (SANIIRI), 12 Dustlik Str, Nukus 742015, Republic of Uzbekistan

Kurbanbaev Sagit
Expert of PIU

Water and Food Security in Central Asia

Viktor A. Dukhovny and Stulina Galina

In Central Asia man learned the art of efficient water use early in history. He used it not only as a means of life-support but also as a natural defender of towns and settlements. He used it as his ally to turn arid steppes and deserts into flourishing oases. Water, in turn, was a great teacher for the farmer, whose patience and diligence allowed him to change this region. An amazing unity of man and nature based on the use of water was formed here.

Throughout history the fight for water was the struggle for survival and water was the basis for developing civilization in this region. It is no coincidence that the need for water resources management has stimulated the development of studies of earth and sky, as well as fundamental mathematics. The names of those who have enriched science such as Al-Fargoni, Al-Khorezmi, Imam Al-Bukhari, Al-Beruni, Ulugbek, Ibn Sino (Avicenna) and many others are well-known worldwide.

Central Asia is a vast region that is a combination of densely populated oases located mainly along the upper and middle reaches of two great rivers and their tributaries, and irrigated areas in their lower reaches and deltas surrounded by deserts, which often swap places due to natural processes that change stream directions or manmade destructive activities.

Most of Central Asia has an arid climate. Scant precipitation (less than 350–400 mm/year), extremely low humidity (22–40% in summer), high evaporation rates (maximum 1,700 t/year), and abundant solar radiation are major climatic features of this region covering an area of more than 300 million ha.

The history of water resources management in Central Asia, an ancient region of our planet where there were the origins of humanity, is of interest now not only from the point of view of understanding the role of water but also for providing progress in the water-consuming sectors and social development. From time immemorial, water and irrigation have been important factors for progress, the development of culture and sciences, and the co-operation of people inhabiting Central Asia. Water

V.A. Dukhovny (✉)
Scientific Information Center of Interstate Commission for Water Coordination (SIC ICWC),
B.11, Karasu-4, Tashkent 700187, Republic of Uzbekistan
e-mail: dukh@icwc-aral.uz

C. Madramootoo and V.A. Dukhovny (eds.), *Water and Food Security in Central Asia*,
NATO Science for Peace and Security Series C: Environmental Security,
DOI 10.1007/978-90-481-9974-7_1,

resources management always required the strict implementation of written or unwritten principles of mutual respect that were often anchored in traditions, rules and customs, as well as in the minds of people who had to appreciate, protect and adore water and everything related to it. According to local traditions, and in accordance with the spirit and moral base of ancient relations between people created initially by the original religions and later by the Muslim religion, water was never a source of profit; water was and must always be the basis for survival and the well-being of humanity.

In Turkestan, since the dawn of time, peoples considered farming as one of the most honorable occupations. The Muslim religious doctrine states that this occupation has divine origins: the first plough was made of a paradisal tree "*tuta*" by the Archangel Gabriel who made a few first furrows and then handed over this plough to Adam. The Shariah calls farmers who themselves cultivate the land as "*ashraf-ul-ashraf*" ("the noblest among the nobles")! Information collected all over the world by the outstanding Soviet scientist Nikolay Vavilov and his pupils allowed them to conclude that the important ancient centers of farming mainly arose in arid and tropical zones as shown in Fig. 1. Central Asia was indicated as one of the six most important areas. Development of these areas was characterized by time shifts and differences in the origin of domesticated plants. B.V. Adrianov wrote that the south-western area of agriculture, which dates from 7,000 to 6,000 years BC, was the most ancient agricultural land in the Middle East and Central Asia. Numerous cultivars (wheat, rye,

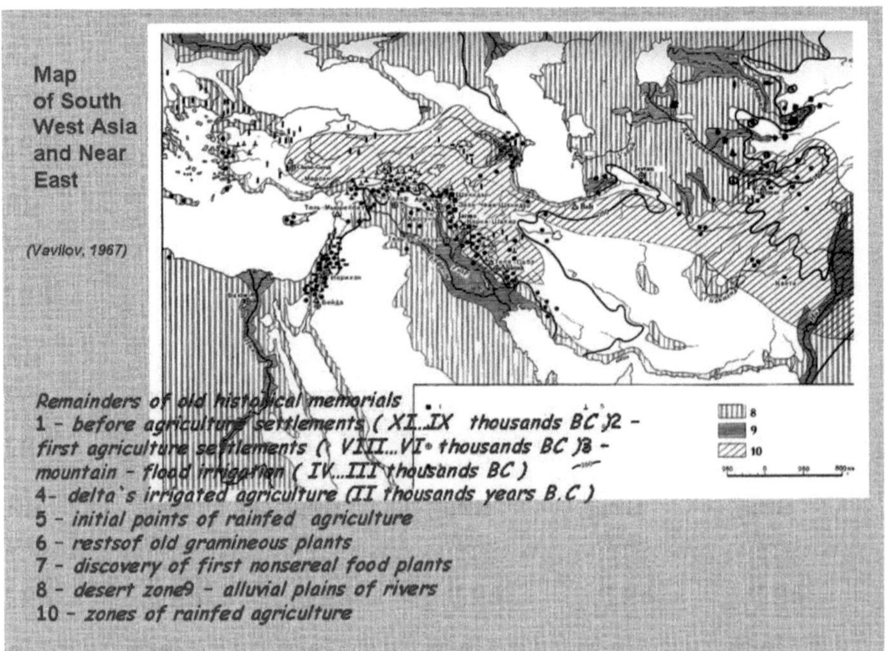

Fig. 1 Important early centers of farming

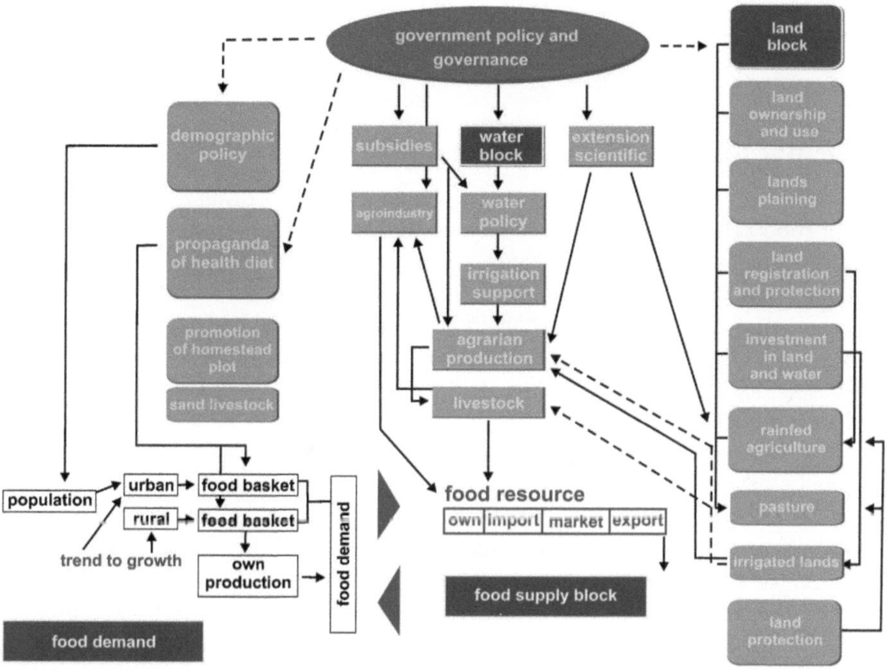

Fig. 2 Balance between food demand and food availability

cucumber, carrot and cotton) were bred here. Many scientists pointed out that the same period corresponds to the development of the ancient centers of civilization in Asia Minor, Mesopotamia, Central Asia and the Mediterranean region.

Sustainable food security of states and the region as a whole depends on major factors that should reflect a common principal: long-term balance between food demand and food availability that needs to be achieved at each stage of future development (Fig. 2). Let's look at these two sides of the problem.

Principal Factors of Food Demand

- *Growth of population* –urban and rural – depending on the demographic policies of government, their care about creating employment, degree of employment in the rural area
- *Food basket* –separately for urban and rural areas that results from traditions, customs, propaganda about a healthy diet
- *Degree of self sufficiency of population* that depends on promotion by state and local authorities of homestead plots (sometimes limited to 0.06 ha) in densely populated areas in Kyrgyzstan, Tajikistan, Uzbekistan, (big promotion in Turkmenistan of plots up to 0.5 ha per family), organization of special garden

plots for the urban population outside of the cities; permission and ability to keep livestock, sheep, goats, chicken, etc.

Principal Factors of Food Availability

As for food availability, these also depend on government policies especially with regards to the principal blocks: lands, water and investment.

Land block includes land ownership, rent and property; size of farms (0.4–2 ha in Kyrgyzstan; 5–30 ha in Kazakhstan and Tajikistan; 3–150 ha in Uzbekistan and a very broad range in Turkmenistan); forms of cooperation; planning of crop patterns (strict state regulations in Tajikistan, Turkmenistan and Uzbekistan); system of land registration and its ability to change and sell (or transfer in subarendy); protection of lands and pasture; state organization (or support) of extension services and agrarian education.

Water block includes water policy; support and development of irrigation; allocation responsibilities between the government and the public private sector (WUAs) for water supply and operation of irrigation networks; water conservation; support and attention to water stability; support and responsibility for reclamation works, etc.

Investment block includes subsidies, tax for lands; payment for water; sponsorship of WMOs and part reclamation of lands; donor involvement and privilege for promotion of effective use of water and lands.

How Do These Factors Look Now?

The rate of population growth was 3.6% in 1980, 2.8% in 1990, and 1.8% in 2000. There is a recent tendency for a population growth rate increase in Turkmenistan, but growth rates remain low in all other states. Nevertheless by the 2030s expect the population of CA to be between 55 and 70 million (Table 1).

Table 1 Population dynamics in Central Asia

	Population growth rate, in % to previous years							Fact (×1,000)	Forecast (×1,000)
CA countries	1999	2002	2003	2004	2005	2006	Avg. (%)	2006	2030
Kazakhstan	101.9	101.5	101.7	101.6	101.9	101.9	101.8	2,956.6	4,536.7
Kyrgyz Republic	101.8	101.3	101.4	101.4	101.3	101.2	101.4	2,943.6	4,109.5
Tajikistan	100.6	102.0	101.3	102.3	101.2	103.9	101.9	7,063.8	11,097.3
Turkmenistan	104.1	103.4	103.3	103.2	103.1	103.0	103.4	6,043.0	13,481.7
Uzbekistan	101.1	101.3	100.8	101.2	101.1	101.3	101.1	26,664.2	34,670.2

A phenomenon that has occurred in the last 5–7 years is a large temporary migration to Kazakhstan and Russia by so called “quest workers”. Most of the migrant workers are from Tajikistan and it is estimated that their numbers range from 0.5 to 0.8 million people or almost 30% of the working population. A similar number of migrant workers are from Uzbekistan but they represent only 5–8% of state labor potential. There are similar migrants from Kyrgyzstan, but an assessment of their numbers is not available. Such migration has created two positive features for food security by reducing the demand for food and increasing the financial capacity of the population. This is particularly important in Tajikistan where the transfer money from migrant workers in 2006, before the financial crisis, was close to 25% GNP.[1]

A study carried out by Matteo Fumagalli “The ‘Food–Energy–Water’ Nexus in Central Asia,” Eucom Policy brief, No 2, October 2008 (Center for European Policy Studies) suggests that the food security situation in Tajikistan is the worst in the region. In the spring of 2008, 1.68 million people in rural areas were food insecure (34% of the rural population) and of those, 540,000 people (11%) were severely food insecure. The situation in urban areas was similar: 500,000 people had a lack of food (33% of urban population) of which 15% or 200,000 had a severe deficit. There is a crisis of the financial capacity of the population and a physical deficit of food. Of all the Central Asian states only Tajikistan has big differences between production and consumption of grain, oils, meat, etc. The European Union through the “Mechanism of collaboration” contributed €314 million to Central Asia in 2007–2010 from which €66 million went to Tajikistan.

Self-sufficiency of rural population for food approached 60% and is therefore significant. The survey of Dr. Stulina “Gender aspects of IWRM in Central Asia and Caucasian” (Tashkent, 2005, GWP, 146 pages) indicates that the portion of food from private home plots (“tamarka”, “melek”) is 50% in Kazakhstan and Kyrgyzstan, 42.3% in Turkmenistan, 36% in Uzbekistan and 22.6% in Tajikistan.

Food basket needed for local population. Table 2 illustrates the changes in principal foods consumed per capita from 1918 to 1980 and compares them to data from a wealthy family at the end of the nineteenth century (Senator Palen, a Russian economist).

Table 2 Average level of food consumption in Central Asia (kg per person per year)

Food	1918	1935	1960	1980	Example of rich people: Senator Palen	Gender survey 2005
Meat	8	13	18	26	32	12
Milk	56	43	110	160	105	130
Grain	190	290	280	240	208	46
Fruits	32	34	36	59	60	180
Vegetables	45	75	90	102	58	105

[1] The situation in 2008 became more difficult as result of price increases of cabbage 281 %, bread 100 %, meat 106 % (USAID Global food insecurity, No 2, 21 May 2008)

Table 3 Provision of population with foodstuff, kg per person per year

		Production		Consumption	
Product	Norm per capita	1990	2003	1990	2003
Bread and bakery	130	92.6	238	170.0	198.0
Meat and meat foods	60	23.6	21.9	32.0	32.3
Milk and dairy products	270	147.9	157.1	210.0	161.0
Vegetables and melons	45.0	138.6	127	107.0	137.0
Potato	76.0	16.4	33.0	29.0	35.0
Grapes, berries, fruits	35.0	68.5	45.6	23.0	44.0
Sugar	14.5	n.a.	9.7	24	n.a.
Vegetable oil		n.a.	8.5	12.6	9.8
Fish		n.a.	0.2	4.9	n.a.
Eggs		120	64.0	120	61.0

Which Food Basket Do We Need?

Table 3 shows the norm per capita of foodstuffs in the food basket during Soviet times as well as production and consumption of the most food secure part of Uzbekistan with a quantity of calories close to 2,800 cal per person per day. The western definition of a food basket with a prevalence of livestock products is very different from the traditional Central Asian basket as the latter demands less water to produce. Deciding which food basket or diet to promote is very important from the point of view of population health, water conservation and the state economy.

What is Our Capacity to Meet the Population's Requirement?

Principal source of increasing food production is the development of irrigation.

- Due to internal and external causes, irrigated agriculture that from time immemorial was the priority of socio-economic development and continues to be the basis for generating livelihood and employment (up to 70% of rural population), has lost, to a considerable degree, its great and obvious economic profitability. A substantial factor was a drop of world prices on agricultural output: rice – 2 times (from 300 to 150 USD/t); wheat – 1.5 times (from 200 to 120 USD/t); and cotton fiber – more than 2 times.[2] Such an economic situation has resulted in the inability of farmers to support the water sector (their current incomes are 100–200 USD/ha instead of 500–1,600 USD/ha in the past). At the same time,

[2] Situation in 2010 changed for the better – increased sharply but it is not clear how this would be reflected on farmers' results.

the social significance of irrigated agriculture that together with related sectors provides employment for 40% of the population (mainly rural), remains topical. Any irregularities in sustainable irrigation water supply caused by deviations from the coordinated schedules of water delivery results in huge social damage (on the verge of catastrophe). Therefore, under current conditions, not only the lack of guarantees for implementing the established procedure of water-sharing by riparian countries, but also the water release regime that is artificially imposed and unacceptable for most stakeholders along with overestimated prices of hydropower production (up to 8.5 cent/kWh) make the existing "order" of water-and-energy exchange of hardly probable.

- The weakening of the economy in these countries and a considerable drop in national per capita income have resulted in a drastic decrease in subsidies and support of the agricultural and water sectors. In addition, procurement of agricultural machinery, fertilizers and other chemicals has declined and water infrastructure, especially at the on-farm level, has deteriorated. As a result, water supplies and land conditions have worsened drastically, affecting the productivity of various crops.
- Introduction of market mechanisms into the agricultural sector (privatization, restructuring large state farms and collective farms into hundreds and even thousands of small private farms) was not accompanied with creating appropriate infrastructure for commodity production and water infrastructure necessary for water distribution and use. As a result, a lot of problems have arisen in servicing new private farmers (consultations, training, and knowledge dissemination) and in procuring agricultural inputs.

Decrease in overall incomes in the irrigated agriculture sector over the whole region (almost two times) and in profitability (a few times) has resulted in impoverishment of the rural population. At the same time agricultural producers are unable to protect their own interests unlike producers of hydropower or fuel that now act in the free market. In 1980, average land profitability in Central Asia amounted to 2,000 USD/ha, compared to 700 USD/ha currently!

Since independence irrigated area has declined in Kazakhstan and Uzbekistan (especially in Karakalpakstan where actual irrigated area decreased by 160,000 ha), remained stable in Kyrgyzstan and Tajikistan, and is growing in Turkmenistan (Table 4). Future increases in cultivated areas are limited by lack of good lands in some areas, a common scarcity of water and a large net cost of new irrigated construction and development. The focus of attention should be on the huge

Table 4 Dynamics of irrigated area (1,000 ha)

CA Countries	1980	1990	2000	2005	2008
Kazakhstan	696	752	769.7	714.3	716
Kyrgyz Republic	422.8	418.5	429.2	410.9	407.6
Tajikistan	670.6	750.5	749.9	763.1	798.9
Turkmenistan	1,080	1,523.3	2,045.7	2,141.8	2,179
Uzbekistan	3,688.1	4,314.7	4,439.2	4,403.9	4,391.8

capacity that we have to increase yields and productivity and reducing the difference between actual and potential land productivity.

Reducing the Gap Between Potential and Actual Productivity

Our approach to achieving potential productivity is based on the assessment of *biological yield* that is not limited by soil, moisture and climatic conditions and characterized only by crop variety.

Potential yield (PY) is the maximum yield that can be achieved in any location with the only limitations being climatic conditions and deviation of weather.

Actual Available Yield (AAY) depends on soil quality, structure, condition of drainage, salinity.

Possible Farm Yield (PFY) is defined by degree of land cultivation, fertilizer use.

Real Yield (RY) is characterized by the degree of on-farm management and organization.

Differences between these yields are estimated in Table 5.

Real yield growth in all states is shown in Figs. 3–7. Grain yield went up significantly in Uzbekistan (wheat ×2; corn ×1.5; time, rice ×1.3), while in Kyrgyzstan yields deviated and even went down (except corn). However yield, particularly of corn, fruits, rice and wheat has a big potential for growth.

The food balance in all states is shown in Table 6. All states except Tajikistan met (or almost met) their needs in grain and had a large amount to export. The production levels of vegetables and meat were similar to consumption levels in all countries, but all have a deficit in oil and sugar.

Table 5 Crop productivity levels (t/ha) in Tashkent province (by S. Nerozin)

Crop	PY	AAY	PFY	RY
Raw cotton	5.8	4.5	3.6	2.3
Cereals	8.8	6.8	5	3.9
Corn	11	8.6	7	3.5
Rice	9.3	7.2	5.9	3.9
Potato	42	33.7	25.2	18.6
Vegetables	39	39.4	26.3	21.9
Cucurbits	45.7	35.4	27.5	16.7
Fruits	31.5	20.8	11.7	4.7
Grapes	34.5	23	12.3	5.5
Forage roots	53	42.5	35	28.9
Kenaf	34	27	22.5	15.5
Perennial grass of past years (hay)	26.5	21.8	16.3	10.4
Perennial grass of past years (green forage)	60	52.5	39.1	27.3
Maize (silage and green forage)	58	45.3	34.8	24.3
Perennial grass of current year (green forage)	42	35.5	27.9	17.5
Perennial grass of current year (hay)	13.8	10.9	8.8	5.5
Annual grass	31.7	26	20.4	13

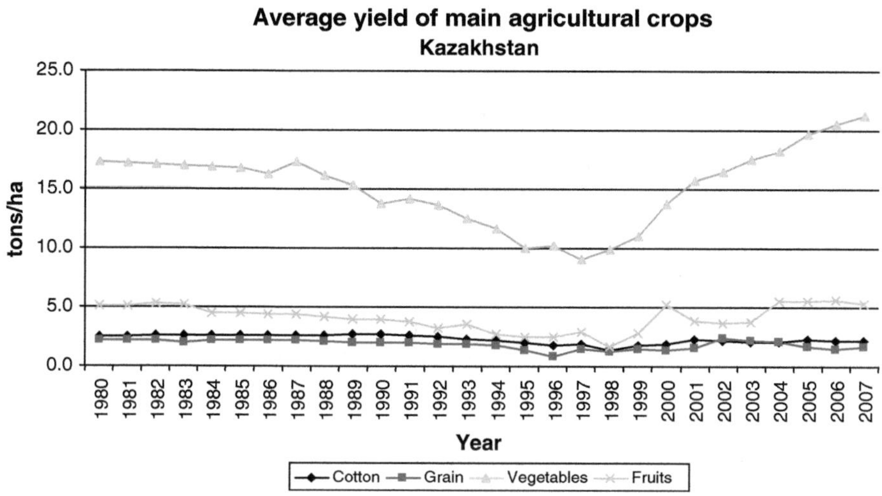

Fig. 3 Average yield of main agricultural crops Kazakhstan

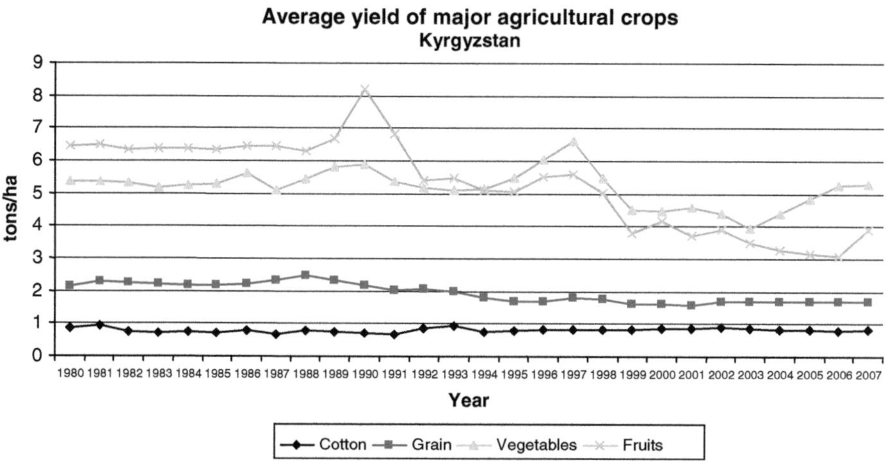

Fig. 4 Average yield of major agricultural crops Kyrgyzstan

What We Can Expect in Future? Our Challenges

Climate change – All scenarios predict an increase in temperature with a resulting growth of water consumption by at least 10–15%.

The most important consequence of climate change is that extreme situations will increase.

The frequency of wet and dry years probability (25% and 75%, respectively) and extreme years (10% and 90%) has increased one, four and two times. Moreover for

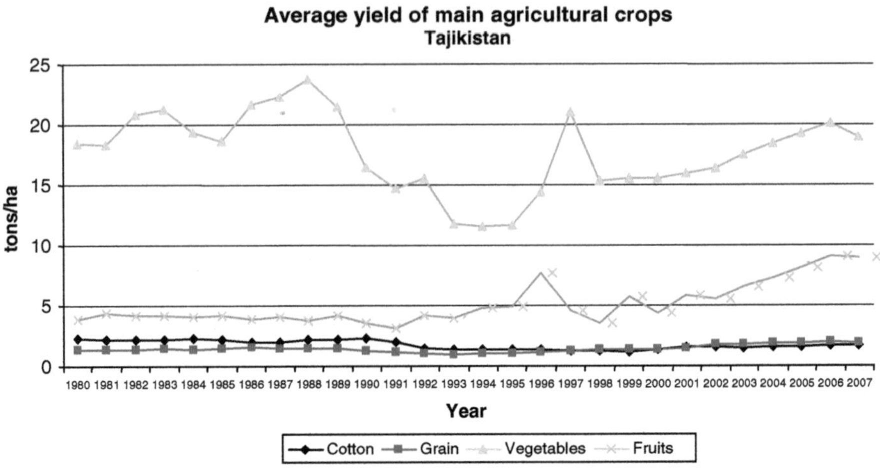

Fig. 5 Average yield of main agricultural crops Tajikistan

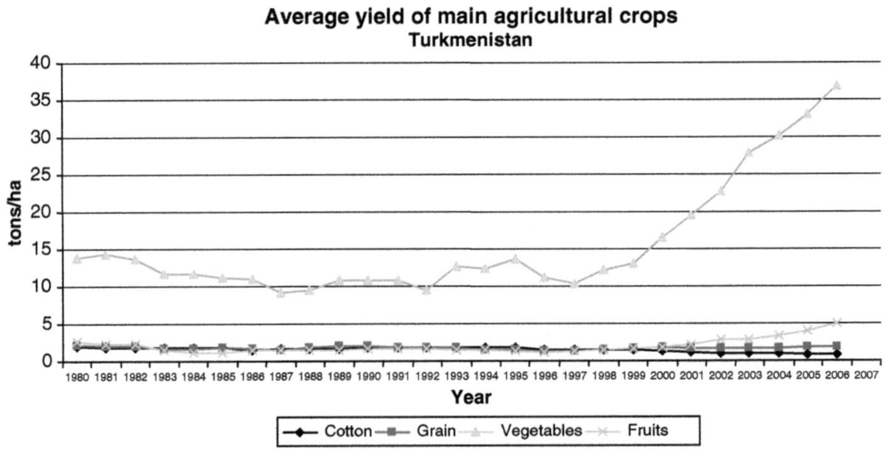

Fig. 6 Average yield of main agricultural crops Turkmenistan

the last 19 years we have had the same number of floods and droughts as for the previous 30 years!

As a result, the role of multiyear regulation and water saving measures grew significantly. However, in reality the region refused to deal with the common interests in multiyear regulation, and problems were created for irrigation as a result of the growing demands for hydropower. An example, the situation in the Syrdarya River with Toktogul reservoir, is described below.

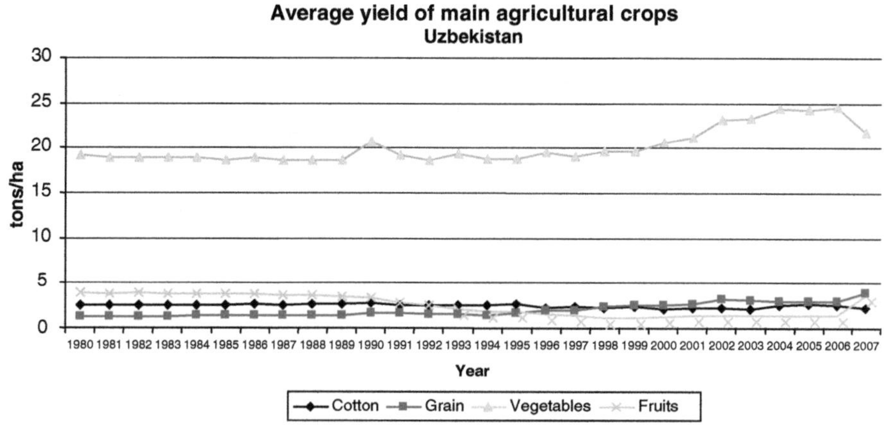

Fig. 7 Average yield of main agricultural crops Uzbekistan

Increasing Importance of Water for Hydropower: Naryn Complex

The reservoir started its operations according to the project operational regime, providing an accumulation of water in winter and appropriate increase in water releases in summer. Over the period of 1982–1990, the average volume of water released during the winter period (October to March) amounted to 3 billion m^3 against 8.1 billion m^3 of summer water releases (Fig. 8). At the same time, inflow into the reservoir made up 2.7 billion m^3 in the dormant season and 9.3 billion m^3 in the growing season, respectively.

Owing to such an operational regime, the Toktogul Reservoir was filled with water up to the design volume by 1988, allowing it to provide the yearly regulation of river flow and accumulating 13–17 billion m^3 of water by the beginning of each growing season. This situation started to change in 1992 with a step-by-step shift from the established operational regime of the Naryn-Syr Darya Reservoirs Cascade and a gradual increase in accumulating water in the summer period and a considerable increase in winter water release. The raising of the rate of winter water release infringes upon the interests of irrigated farming and other water consumers including ecosystems (Fig. 9). Since 1994, this tendency has turned into a principle of management, and as a result, the reservoir is being operated according to a regime necessary for energy production with abrupt reductions in the volumes of water released for irrigation. Volumes of winter water releases have increased up to an extreme amount of 9.7 billion m^3 during the autumn–winter period of 2007–2008, and the volume of summer water release was reduced to 3.6 billion m^3 during the dry years of 2001 and 2002.

Figure 10 clearly shows that over the period 1981–1991, the Toktogul Reservoir was operated according to the design regime of accumulating water for the purpose of over-year regulation (especially accumulating water from 1986 to 1988 with abundant water inflows) and facilitated almost smooth overcoming of the

Table 6 Production versus consumption of food in Central Asia (1,000 t)

	Grain		Oils		Vegetables		Meat		Total product	
	Product	Consump.	Product.	Consump.	Product.	Consump.	Product.	Consump.	Export	Import
Kazakhstan	15,446	10,644	88	183	2,254	2,269	673	737	692.7	933.5
Kyrgyzstan	1,707	1,858	29	40	1,240	1,236	198	201	113.3	102.4
Tajikistan	661	1,028	30	55	379	381	36	55	204.1	110.5
Turkmenistan	2,295	2,340	29	52	140	141	193	200	86.4	99.4
Uzbekistan	5,147	5,607	289	303	783	788	518	527	834.7	166.4

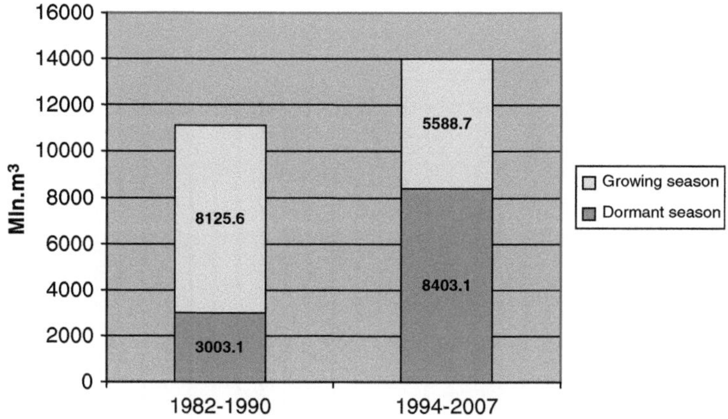

Fig. 8 Average volumes of water releases from the Toktogul Reservoir over the periods of 1982–1990 and 1994–2007

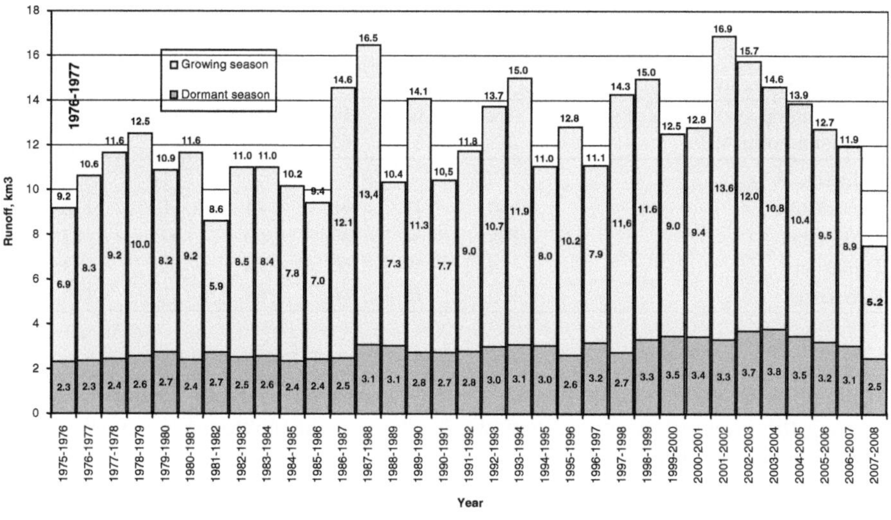

Fig. 9 Inflow into the Toktogul reservoir over the period of 1975–2008

consequences of droughts. It is necessary to note that in spite of relatively low water availability over the period of 1988–2002, it became possible, thanks to the well coordinated work of water organizations in the region and the BWO "Syr Darya", to avoid reduction of water supplies to the irrigated farming sector and infringing interests of all water users even in dry years (2000 and 2001), when the level of water availability in the lower reaches of the basin dropped by up to 50%. However, the following period of wet years (2002–2006) was not used for accumulating water resources as envisaged by the Operational Rules for the Naryn-Syr Darya Reservoirs Cascade. As a result, droughts in 2007 and 2008 created a catastrophic

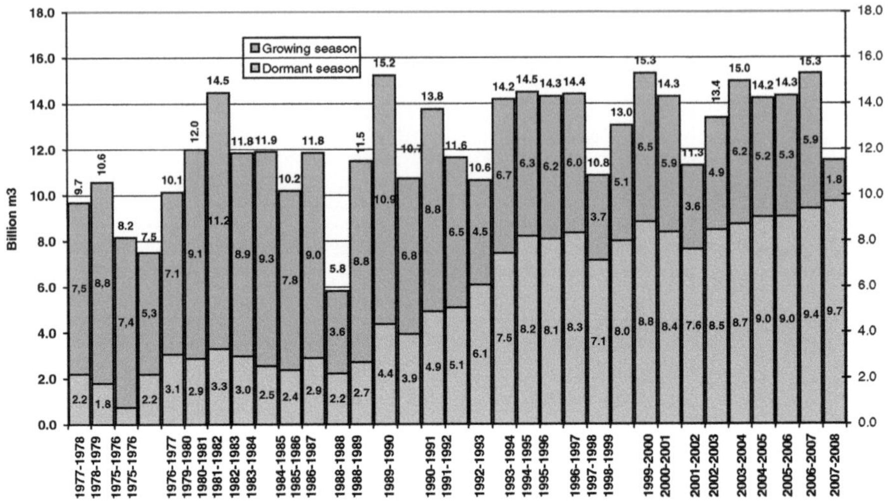

Fig. 10 Water releases from the Toktogul reservoir

Table 7 Water availability in Syrdarya (2008)

Water natural availability of mean long-term norm, %	April 77	May 86	June 65	July 55	August 76	September 70
Delivery, %						
Kazakhstan	150	147	86	44	58	178
Kyrgyzstan	105	64	57	60	67	81
Tajikistan	34	59	69	74	85	81
Uzbekistan	120	76	60	58	72	105

situation with water availability in the Syr Darya River Basin because prior to starting the growing season, drawdown of water in the Toktogul Reservoir practically reached the dead storage level.

There has been a retreat from the key requirement of operational rules of the Toktogul Reservoir. A shift has occurred from over-year regulation towards seasonal regulation for the benefit of energy production with the result that the reservoir is filled by the beginning of the autumn–winter season and is emptied by the beginning of the growing season. This regime has had two consequences: (1) a lack of irrigation water availability in summer and (2) artificial floods during winter. The multiyear regulation of water was destroyed by actions in favor of hydropower and in 2007–2008 did not assist in water availability. Moreover, if natural water availability in 2008 was 75%, real water availability in Syrdarya deviated from 62% to 75%! (Table 7)

The reason for this shift is very clear: commercialization of hydropower and the growth in energy prices caused the increase in winter water release for power as opposed to meeting irrigation requirements in the summer. As result the additional benefits in energy were increased by 30 million USD and caused additional losses in irrigation of 120 million USD.

In recent years, one more factor of disorganization has arisen: the practice of managing the Naryn Hydropower Stations Cascade by the organization "Kyrgyzenergo". In the guise of need for daily regulation of energy production, after 6 p.m. this organization sharply decreases the volume of water released from the last unit of the cascade (the Uchkurgan Hydropower Station) into the riverbed at the entrance to the Fergana Valley, although this type of daily regulation could be implemented at upstream hydropower stations of this cascade.

In 2005, international experts noted inadmissible fluctuations of water levels in the upstream pool of the Uchkurgan hydro scheme caused by daily fluctuations of water releases with amplitude of flow rates $\pm$ 200 m^3/s according to the energy consumption schedule at the Uchkurgan Hydropower Station. Last year, this phenomenon had a catastrophic effect, since every night the river flow was completely blocked because, according to Kyrgyz managers, of the lack of need for electric energy during night. This operational regime of the Naryn Hydropower Stations Cascade considerably troubles water diversion to the systems of Big Fergana Canal and Northern Fergana Canal that supply water for irrigation of agricultural lands in the Fergana Valley. The water level in the river varies from 0.5 to 2.5 m over 1 to 3 h (Fig. 11), and results in lowering the extent of sustainable water supply through irrigation canals and in infringing on the design regulations developed for operating the hydraulic structures. Every day, over an area of 350,000 ha in the Fergana Valley, farmers and other water users have to overcome the consequences of a poorly-regulated water supply, resulting in a drastic reduction of efficient water use!

It should be noted that such uncertainty plays against water supply for food production guaranteed in accordance with Article 11 of the International Covenant on Economic, Social and Cultural Rights and the follow-up comments of the Committee on Economic, Social and Cultural Rights (Comment 15) (13). In principle, the approach with energy supply in winter in exchange for additional water releases in summer can be employed; however, it is necessary to exclude the

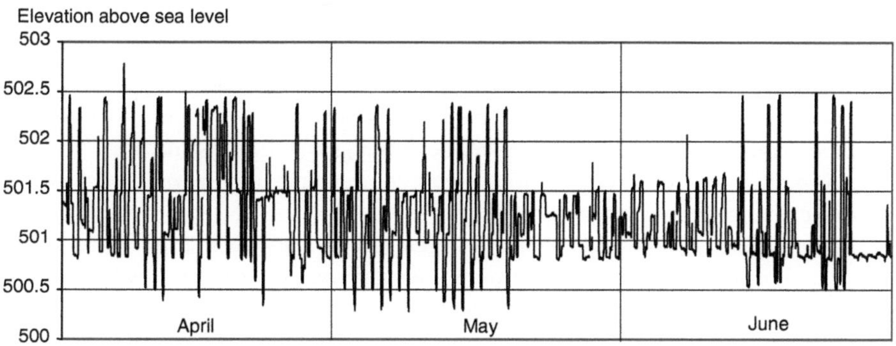

Fig. 11 Water level fluctuations in the Naryn River downstream from HPS cascade, since the beginning of the growing season in 2008

discussion of prices of electric energy generated and supplied in summer and their linking to prices of fuel. *Electric energy generated in summer in excess of demands of Kyrgyzstan and for the benefit of irrigation should be compensated for in winter also by electric energy rather than by natural gas or coal*; at that, its price has to be identical, although, in principle, summer electric energy has undoubtedly a lower price in the Central Asian market as it is a less scarce resource. In this case, all speculative approaches will be excluded.

How We Can Ensure Future Food Security?

Some principal directions:

1. Development of Integrated Water Resource Management (IWRM).
2. Establishment of a strong state policy oriented towards food security.
3. Increasing water rights for irrigated agriculture at all levels of water hierarchy.
4. Strengthening collaboration between states for establishment of a guaranteed regime of water delivery.

The Implementation of IWRM in Central Asia Started in 2002

IWRM is based on the following key principles that define its practical backbone:

- Water resources management is implemented within the hydrological units in concordance with geomorphology of the drainage basin under consideration
- Management takes into consideration assessment and use of all kinds of water resources (surface water, ground water, and return water) and the climatic features of the regions
- Close co-ordination of all kinds of water uses and organizations involved in water resources management, including cross-sectoral (horizontal) co-ordination and co-ordination of hierarchical levels of water governance (basin, sub-basin, irrigation system, WUA, with the farm as the end user)
- Public participation not only in the water management process, but also in financing, planning, maintaining and developing water infrastructure
- Setting the priorities of ecosystem water requirements into the practice of water management organization
- Participation of water management organizations and water users in activities related to water saving and control of unproductive water losses; water demand control along with resources management
- Information exchange, openness and transparency of the water resources management system and
- Economic and financial sustainability of water management organizations

The IWRM conception was coordinated and approved by all water authorities in Uzbekistan, Kyrgyzstan, and Tajikistan in May 2003. A comprehensive approach for social mobilization was developed along with a training program for social mobilization and capacity building at the levels of WUAs and irrigation canals. Regular training seminars and sociological surveys conducted in the frame of the project provide new opportunities for involving all stakeholders in reforming the water sector in the Fergana Valley. Thanks to project efforts, new water user associations were established, and earlier established WUAs were restructured. Since July 2002, the project conducted monthly planned (according to the project agenda) and unplanned training seminars for specialists of water management organizations, water users and members of NGO's in the Fergana Valley. Particular attention was paid to dissemination of the IWRM ideology. A communication network based on e-mail and linked to all key project participants (the SIC ICWC, national departments, provincial water management organizations, pilot canal administration, WUAs) was developed. The project has established an on-line Information Management System (consisting of a database, a set of mathematical models, and GIS), which is a powerful tool for planning, operational analysis, and improving the water allocation process and actual water distribution.

Alternative organizational structures for water management at the level of main irrigation canals were created as new departments called Canal Administrations (on the Aravan-Akbura Canal in Osh Province in Kyrgyzstan, Khodja-Bakirgan Canal in Soghd Province in Tajikistan, and South-Fergana Canal in Uzbekistan). In December 2003 the Pilot Canal Water Users Unions were established and officially registered on all pilot canals; and the joint governance principle was put in practice. The agreements related to joint water governance were signed, and the Canal Water Committees consisting of representatives of superior state water management organization (WMO) and water users (CWUU) were created.

The first steps towards establishing the procedures of water resources planning, record-keeping, reporting and monitoring at each level of the new water management hierarchy were made. An effectual factor of transition towards IWRM is participation of representatives of civil society in the governance process that is also legally fixed. It is expected that activity will be implemented at all levels of the water hierarchy. Many technical aspects also depend on public participation. It is not an easy task to provide guaranteed and equitable water distribution over the entire irrigation system. When water of sufficient quality is delivered in line with planned amounts, increases in productivity of water and land resources may be expected. Water users should participate with more precise specifications of command areas for each irrigation canal, assessment of their water demands, and accounting for additional available water sources (ground water, return water). Adjusting water supply, rotation and use depending on weather and economic conditions, as well as improving hydraulic measurements and record-keeping at all levels of the water management system are also their functions. To tackle issues as they arise, it is necessary to establish extension services that assist water users in the introduction of new technologies, advanced practice of planning and production, and solving water distribution problems. The project has developed

"The Model Regulations on Canal Water Committees", as well as recommendations for their adaptation on each of three pilot canals.

The project has rendered technical assistance in inspections and additional equipping of flow-measuring structures on pilot irrigation canals (an enormous program was implemented to establish water-metering systems within pilot WUAs). This activity allowed the establishment of proper water record-keeping on the pilot canals and within WUAs, making the water distribution process more transparent. The project has started real-time management of the water delivery process on pilot irrigation canals and within pilot WUAs in the form of monitoring and updating the planned water supply schedule based on water user applications taking into account weather conditions during a growing season. This is the first step towards equitable and rightful water distribution and, at the same time, an attempt to reduce unproductive water losses.

Preparing the passports of demonstration fields within pilot farms allowed the creation of an instrument for analyzing farmers' production reserves and potential with the purpose of improving productivity of land and water resources. Testing the instrument for forecasting water consumption in line with weather conditions is conducted in the real-time regime, and its introduction at a broad scale during the next phase of the project is planned. Our analysis shows that on nine of ten pilot plots, the land and water productivity was perceptibly improved. On one pilot plot located on the SFC, where the farmer did not follow the project recommendations, productivity has declined. Many women were involved in discussions related to land management and water productivity and other water resource management problems in the Fergana Valley. Based on outcomes of these activities, an enabling environment was created for the wide introduction of extension services for farmers in the Fergana Valley.

Introducing IWRM has shown that the involvement of water users (community initiative) allows considerably improvement in the efficiency of water use and a decreased head water withdrawal by more than 25%, creating a system of fair and equitable water use, which, by its nature, is close to traditional water use in accordance with the Sharia canons (Figs. 12 and 13).

This experience was applied to upscaling in Uzbekistan on seven rayons of different provinces, same as in Tajikistan and other states. If we can achieve reduced water delivery, even of 20%, on the basis of this experience, significant growth of food production, well being, and income in rural areas and a reduction of poverty should occur.

The agricultural and water strategy of the state should be a framework to support the efficiency of IWRM and simultaneously promote the growth of food production. It needs to create concrete conditions for the development of private initiatives and transform the farmer to become a real owner of his production with a long-term commitment through the input of his own skills.

The average ratio of production between cooperative and private farms approached 1.5 and indicated correctness of this line of thinking (Table 8).

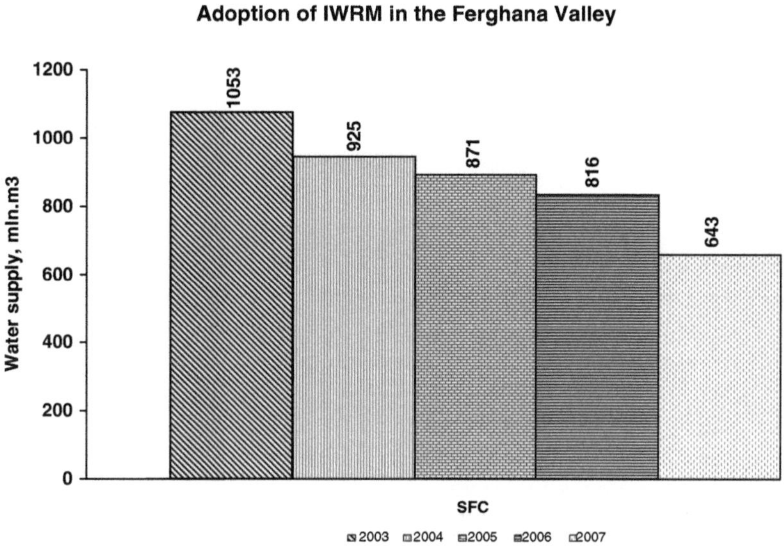

Fig. 12 Reducing the water diversion into the South Fergana Canal (SFC) due to introduction of IWRM in irrigation season

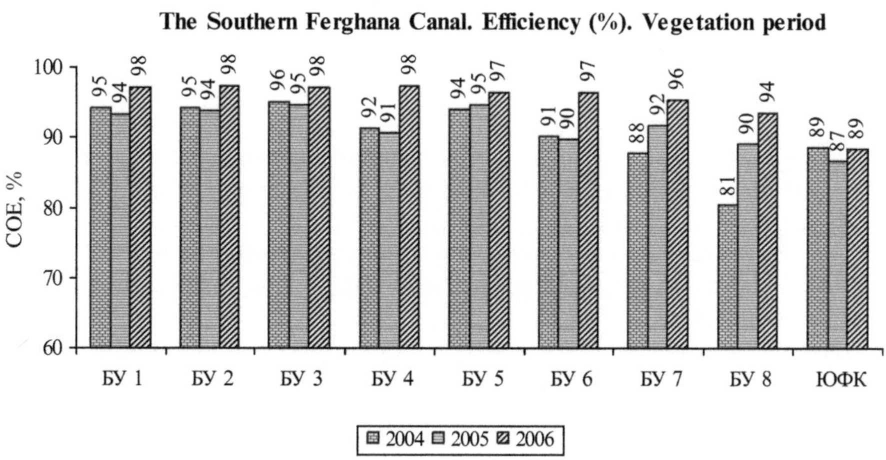

Fig. 13 Dynamics of the efficiency at water-balance sites and the SFC as a whole

Other Measures Required from Governments

- Creation of climate for strengthening WUAs and their transformation to multi-sided unions

Table 8 Relation between yield of land in cooperative and private farms, t/ha on 2003

Crop	Private farms	Cooperative farms	Ratio
Corn	3.84	4.6	1.19
Vegetable	22.48	30.3	1.34
Grape	2.38	4.9	2.05
Fruits	3.21	5.7	1.77
Potato	21.10	29.9	1.41

- Establishment of a correct system of financial interrelation "farmer-WUAs-processing-marketing" that will support stability and insurance of farmers and their partner in the future
- Finalize "market approach", reduce (in Tajikistan, Turkmenistan, Uzbekistan) strong state planning to no more 50% of lands with free ability of farmer to select crop, traders and conditions
- Support creation of market oriented rural infrastructure for assistance that will be in close cooperation with farmers and their WUAs

Strengthening of water governance is a huge priority of global, regional and local communities. Each citizen of the planet is a water user, and all those who belong to the water community and service are servants of God! It is most important to bring up these issues on the global level. There should be preparation of a global water security charter that includes some major principles:

1. Water is a natural and common social resource that should be first of all used to meet:
   (a) Drinking and communal needs
   (b) Requirements of food production (irrigated farming) and industrial sectors
   (c) Demands related to health, nutrition, minimal employment, and well-being of the population, especially of its most vulnerable groups and
   (d) Environment needs
2. Water must be treated as a commodity only after special processing that turns water into a consumer good (bottling, desalinization, etc.) or if it can be replaced by its virtual form
3. Equitable and reasonable access to water for everyone and water use in amounts that correspond to the most advanced technologies are moral and legal rights that should be secured by governments
4. Nobody can possess the preferential right of using water resources of transboundary watercourses; at the same time, diversion of water from transboundary watercourses can be made only in the amounts that are agreed upon and justified based on modern technologies of water delivery and use
5. Nobody possesses the right to change the flow regime of a river in such a manner that results in man-made droughts or floods in downstream reaches
6. Any actions on transboundary waters should not inflict significant damage on other stakeholders
7. GWG should be structured as a robust set of legal and social rules and norms that cover all levels of water hierarchy and all branches of public life including the

economy that, in its turn, will provide a guarantee of water availability for nature and human needs

Based on above considerations, we propose to consider strengthening the system of *water governance* by means of elaborating integrated measures that encompass legal, institutional, financial, and social components with the purpose of co-ordination of all levels of the water management hierarchy. These measures should remove all weaknesses and vulnerability of the existing system of water governance, and also opportunities for ignoring the interests of other water users or creating the uncertainty in their water use regardless of a level of the water management hierarchy.

Strengthening *water governance* at the transboundary level (international watercourses) is especially important. Available documents that form the international water law related to non-navigation use of transboundary watercourses (the Helsinki rules, UNECE Convention on the Protection and Use of Transboundary Watercourses and International Lakes) envisage the rather correct and fair principles of water use; however, their practical application (even in case of their ratification by all parties) is quite difficult owing to a number of provisions that can be interpreted by each Party of these conventions in its own interests. For example, these are the following provisions: (i) equitable and reasonable water use; (ii) preventing considerable (tangible) damage; and (iii) co-ordination of actions that can change quantitative and qualitative characteristics of water bodies. One more problem is the definition of boundaries of national sovereignty, the rights, duties, and powers of riparian and basin organizations, as well as the provision regarding obligatory collaboration, etc. All these ambiguities, shortcomings and vague wordings were repeatedly mentioned in numerous studies pointing to the fact that compared with the Helsinki Rules there are more ambiguities (for example, such provisions as "reasonable water use," "preferential use," "situations threatening to human health or deterioration of social and economic conditions," "ecological flows," etc.).[3]

Therefore, one of topical tasks is the preparation and approval of detailed recommendations (if not revision of these documents) or the protocol concerning the key principles of UNECE Convention, which will help the riparian organizations to be clearly guided by their rights and duties in all aspects of the international water law. An important matter of transboundary water resources management is the establishment and proper operation of basin organizations. In this case, more clear and specific rules and regulations rather than simply recommendations regarding the establishing of joint basin organizations and possible spheres of their activity[4] are also required. The UNECE has already attempted to elaborate such rules for the river basins in Eastern Europe, Caucasus and Central Asia[5] based on the generalization

[3] S. Bogdanovich, International Law of Water Resources, Contribution of the International Law Association, Kluwer Law International, London, 2001, 436 pp.

[4] Ch. Gopalakrishnan, C. Tortajada, A. Biswas "Water institutions, political, performance and prospects", Springer, 2005, 209 pp.

[5] The UNECE, River basin complexes and a new institutional mechanism in the field of transboundary water collaboration. 2007.

of the international experience. This document should include the rules of establishing the basin organizations depending on their objectives and the field of activity, as well as the provisions concerning their institutional framework, financing, etc. The important constituents of this document are the provisions concerning the procedures of arbitrage, evaluation of damages and their compensation (these functions should be implemented by basin organizations).

Water rights should be clearly specified at all levels of the water management hierarchy. Ensuring water rights includes not only their declaration and establishing the procedures for water allocation (a proportional share of available water resources, quotas or the certain order of priorities) but also enabling an environment for observance of these rights that embraces the following components:

- Sustainability of water supplies at all levels of the water management hierarchy, and first of all, at the level of transboundary water management (it is impossible to speak about sustainability if upstream riparian organizations regularly change the regime of water releases from the reservoirs located on their territories)
- Physical infrastructure and agreed upon procedures for multi-year regulation of river runoff and its control
- Procedures for adjusting national water quotas
- Physical infrastructure and the agreed procedures for stable and uniform water distribution among water users and water consumers
- Participatory water resources management
- Mechanisms for supporting sustainable operation of local water organizations
- Potential for mitigating water scarcity due to effects of destabilizing factors
- Financial and material well-being of the water sector

There should be a special emphasis on supplies of water to ecosystems since most countries take into consideration the needs of the natural complex according to the leftover principle. However, satisfying the hydro-ecological requirements of ecosystems is very important for supporting their economic function; and disregarding these requirements can result in the most destructive consequences for mankind in the future.

The following aspects play a very important role at the national level:

- Establishing the National Water Council as a single body that integrates the participation of all substantial stakeholders in water resources management and their perspective development
- Elaborating the National Water Strategy and its co-ordination with all executors
- Regular monitoring and evaluation of all destabilizing factors and adaptation to their changes (or mitigating adverse affects)
- Protection of the rights of the underprivileged when transferring water supply services to private companies
- A role of the state in maintaining the water infrastructure (water supply and sanitation, irrigation and drainage) and
- Liability for violation of the rights of some economic players in access to water resources

Table 9 Situation with irrigated lands on the NIS and Eastern Europe (1,000 ha)

	1990	2004	Actually irrigated
Russia	5,799	3,506	2,600
Ukraine	2,455	1,100	700
Uzbekistan	3,908	4,230	3,960
Kazakhstan	2,160	1,290	1,060
Turkmenistan	1,240	1,760	1,700
Bulgaria	1,250	40.0	
Czechia	133	10.0	
Germany	500	200.0	
Hungary	300	100.0	
Poland	301.5	83.3	
Romania	3,205	850	500

Last but Not Least: Attention to World-Wide Irrigation and Drainage

It is well known that irrigation and drainage are responsible for the creation of almost 50% of agrarian production on the global scale. Nevertheless over the last 25 years this sector has lost the attention, investment and government involvement in support of this most important source of food production. One indicator: the failure of irrigated lands in NIS and Eastern Europe states. As can be seen in Table 9, almost 10 million ha of irrigated land were lost from global potential or almost 4% of global irrigated lands. Who assesses the consequences of this on global food security?

Comparison of *food self sufficiency* and *virtual water* approach. Referring to my statement in "Water and globalization in Central Asia" (Irrigation and Drainage, 56, 489–507, 2007), I need again to underline that assessment of the water deficit based on virtual water can destroy the real ability of the state for national food sufficiency. Take into account the deviation of food prices on the global market security. Developing states, through their own farm production, generate food security and solve nutrition problems which are of social importance. In addition, irrigated agriculture provides employment and income, not only in direct agriculture, but on all associated sectors, service, etc. We must remember that the cost of our working place in industry is 8–15 times more than one place in irrigated agriculture! So, let us protect the role of our sectors!

Global Challenges in Water Management: The Canadian Context

Chandra Madramootoo, Robert Baker, Aly Shady, and Helen Fyles

World Land Use and Irrigation

One third of the land on Earth is suitable for growing crops and livestock grazing and one third is in forest with poor soils. About 1,369 million ha or 10% of agricultural land is used to grow annual crops and of this, approximately 270 million ha are irrigated. The majority of the world's irrigated area (70%) is in Asia where China and India each have approximately 55 million ha of irrigated area (Table 1). Most of the world's irrigation is for rice, a wetland crop in Asia. Most other grain crops are grown in semi-arid climates and no food crops are grown in tropical forests. Negligible food crops are grown in wetlands.

Challenges for World Food Security

The number of hungry people rose from 850 million in 2005 to 925 million in 2008 to over one billion in 2009. The causes underlying this sharp increase in food insecurity include relatively recent events such as conversion of land to biofuel production, rising oil prices, lower cereal reserves, trade speculation, and droughts in key grain producing countries all of which contributed to higher food prices. The 2008 financial crisis further exacerbated global food security due to a sharp decline in remittances to developing countries from people working abroad as jobs losses in developed countries increased. There are also long term issues underlying global food insecurity including the growing world population, changing food consumption patterns, international trade policies and reduced national and international investments in the agricultural sector and increasing water scarcity in many regions of the world.

C. Madramootoo (✉)
Faculty of Agricultural and Environmental Sciences,
McGill University, Macdonald Campus, Ste-Anne-de-Bellevue, Quebec, Canada
e-mail: chandra.madramootoo@mcgill.ca

C. Madramootoo and V.A. Dukhovny (eds.), *Water and Food Security in Central Asia*,
NATO Science for Peace and Security Series C: Environmental Security,
DOI 10.1007/978-90-481-9974-7_2,

Table 1 World land use (FAOSTAT 2004)

	Land under irrigation (1,000 ha)	Share of world total, %	Share of cropland that is irrigated, %
Africa	12,879	5	6
Asia	193,869	70	33
Central and North America	31,408	11	12
Europe	25,220	9	8
South America	10,499	4	8
Oceania	2,844	1	5
World	276,719	100	18

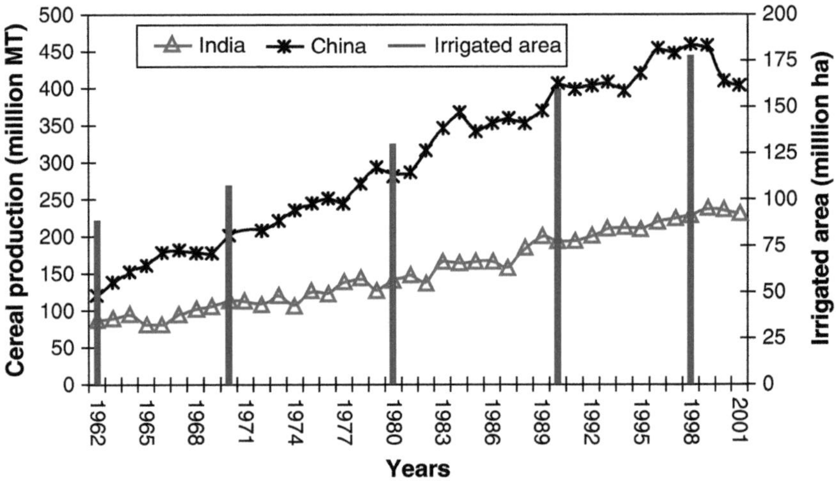

Fig. 1 Cereal production and irrigation development in Asia following the green revolution

Countries with the highest proportion of hungry people are predominantly found in Sub-Saharan Africa, but parts of Asia, and Central and South America also have areas where over 25% of their populations are undernourished. Malnutrition tends to be high in semiarid and dry subhumid climates subject to variable rainfall, dry spells and droughts.

Food insecurity is not a new global issue. Severe droughts and famines in Africa, Asia and Latin America in the 1940s and 1950s occurred at a time when populations were beginning to grow sharply and many countries invested heavily in irrigation to combat malnutrition and death due to hunger. This was the beginning of the 'Green Revolution' with a major emphasis on raising crop yields through genetic improvement and increased inputs such as fertilizers, pesticides and water. Irrigated area has expanded significantly since that time (Fig. 1).

Irrigation made it possible to move from one to two or even three crops of rice per year. Yields of mainly irrigated cereals such as rice and maize have more than doubled since 1960, and those of wheat increased threefold while the predominantly rainfed crops such as sorghum and millet recorded only 30–60% increases in the yield index (World Bank 2006). Irrigation also has the potential to stabilize crop production, improve crop quality and allow for diversification in farm production including the introduction of new crops that are not generally viable under dryland farming.

On average, agriculture accounts for 70% of the world's water use and there is an intrinsic relationship between a country's renewable water resources and its capacity to produce food for domestic consumption and export. In water-scarce countries, an increasing amount of food, particularly cereal grains, are imported to meet local requirements for food and relieve demands on scarce water resources. However, worldwide, 92% of world cereal production is consumed where it is produced leaving only 8% available for trade. Demand for this limited amount of cereal will increase as water-scarce countries become more dependent on imported food.

To increase agricultural production and reduce input costs a number of tools are in use today that were not available in the 1960s. New and more effective ways to use irrigation water have been developed and reduce the amount of water needed to provide the same or even higher crops yields. Biotechnology to increase crop yields and disease resistance are being studied worldwide and bioengineered crops have been adopted by 23 countries. Solar power is increasingly used as a source of energy to replace high cost fuels. Solar power runs pumps, pivots and drip irrigation and is particularly valuable in the developing world where access to, and the cost of, fuels is a severe limitation to the small farmer.

Challenges to World Water Supply

It is estimated that 1.2 billion people lack access to clean drinking water, 2–3 billion lack adequate sanitation, 4 billion are without adequate sewerage services, and 5–10 million deaths per year are related to poor water quality. Freshwater per capita has declined worldwide from 12,050 m^3 in 1950 to 7,310 m^3 in 2000 and is expected to reach 4,580 m^3 in 2050. Water scarcity is not evenly shared between or even within countries. For example, the average per capita water availability in China was estimated at 3,400 m^3/year in 2005 (UNESCAP 2006) while in the Yellow River Basin, the number drops to less than 600 m^3/year (Giordano et al. 2004). Regions that will be particularly affected by water scarcity in the future are in Africa and Asia (Fig. 2).

Rising populations, increased urbanization and higher calorie diets that include more meat and dairy products, all have a critical effect on the demand for fresh water. The world population is projected to reach 7 billion in late 2011, up from the current 6.8 billion, and surpass 9 billion people by 2050 (United Nations 2009a). More than half of the world's population currently live in urban areas and this is expected to increase to almost 60% by 2030 (United Nations 2009b). As the global urban population grows, more water will be required worldwide for domestic and industrial use and water demands will become concentrated in relatively small areas. There is already intense competition for water in peri-urban zones in

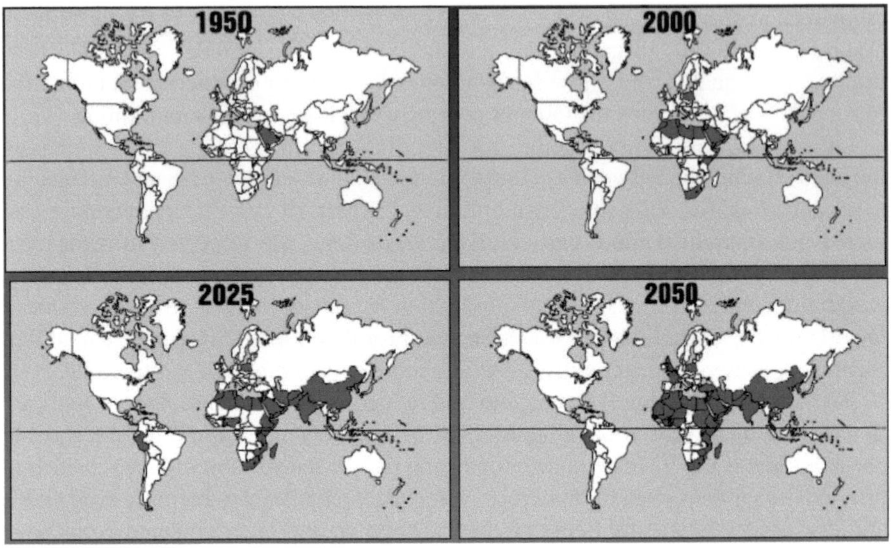

Fig. 2 Evolution of water shortage in 100 years

developing countries. The consumption of more meat, processed foods, and higher value food often rich in sugar and oils require more water to produce and consumptive use of water for food supply is significantly higher where there is development and improved lifestyles (SIWI/IWMI 2008).

Environmental problems such as salinity, flooding and degradation of downstream water quality or overuse of surface waters leading to loss of riverine ecosystem biodiversity that were allowed to happen in the past are being addressed. Modern environmental legislation now acknowledges the importance of aquatic ecosystems and requires that adequate water flow be maintained in rivers and streams, creating further demands on freshwater supply.

Climate change will further aggravate the availability of freshwater resources particularly in semi-arid and arid areas. Glacier melt and reduced snow pack in recent decades is expected to accelerate, resulting in reduced long term water availability and earlier spring runoff in regions supplied by melt water from major mountain ranges (IPCC 2007). Reduced rainfall and higher evapotranspiration is projected for many semi-arid areas, and drought-affected areas are expected to increase in extent. Higher sea levels are anticipated and may result in flooding and contamination of freshwater aquifers in coastal areas.

Addressing Future Water Needs

Increased global demands for water, spread unevenly around the world and across the landscape will require improved water storage, cooperation between countries sharing water resources, interbasin transfer and serious efforts to conserve and use water more effectively. The significant consumption of water by agriculture means that this sector must play a major role in reducing the non-beneficial consumption of water for crop production.

Dams are an essential part of our water storage infrastructure. More than two thirds of world water reservoirs are used for irrigation and food production. For almost 5,000 years dams have been used successfully to collect and store water and manage discharges to provide the large quantities of water to sustain life and support growth and development. Dams are a tool for water management, irrigation, flood control and hydropower. Water storage at existing dams could be reallocated to provide additional water supplies (Fig. 3).

Moving water from water rich to water poor areas or interbasin water transfer may be necessary in some countries. For example, in China water is unevenly distributed between the northern and southern regions. The south has 81% of the water resources, 53% of the population and 35% of the cultivated land while the north has 19% of the water resources, 47% of the population and 65% of the cultivated land. Transferring water from the south to the north would allow for increased overall food production. The alternative would be to move people from the water poor areas to water rich areas.

Transboundary cooperation of countries which share water resources would result in an overall higher food security and more effective use of water resources. Participation in transboundary water development can contribute effectively to economic growth, poverty reduction, improved health and nutrition and the promotion of peace and security. For example, the Nile River flows through ten riparian states with a total population of about 250 million people. Five of these states are among the ten poorest in the world. The basin is about 3 million km^2 and contains a lake area of 81,500 km^2. The length of rivers and tributaries is 37,500 km and an area

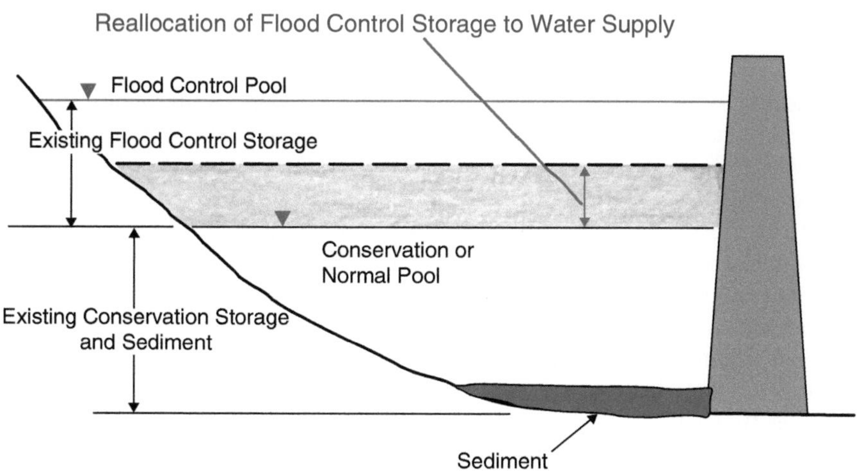

Fig. 3 Reallocation of storage at existing dams to optimize project benefits

of swamps of 70,000 km^2. The current challenges in the Nile Basin are: extreme poverty, instability, rapid population growth, environmental degradation, natural disasters (floods, droughts, etc.), complicated hydrology of the basin, low specific yield, equitable use and no harm principle. Opportunities exist however as the river is relatively undeveloped in its upper reaches and the potential is great for water saving, agriculture, and sharing of hydro power. Serious steps for cooperation have already been taken and are an incentive for donors. Future prospects for the Nile Basin include:

- Build trust and confidence between governments
- Strengthen the indigenous capacity of each region
- Take advantage of new information technology
- Policy reforms, legal and institutional overhaul
- Emergence of civil society and their active participation
- Long term commitment, vision and political will
- Facilitation and support by external support agencies

Development and cooperation potential:

- High hydropower generation potential: shared grid
- High irrigation potential: meet all of Africa's future food needs
- Improved river navigation: trade and transport
- Improved water quality: better health and more fresh water fisheries
- Ecological conservation and stewardship
- Poverty reduction –economic growth

Water Management in Canada

Water Resources and Use

Canada has 7% of the world's renewable water supply and is the country with the third most renewable freshwater. Only Brazil and Russia have more. Canada also has about 25% of the world's wetlands – the largest wetland area in the world. Almost 9% or 891,000 km^2 of Canada's total area is covered by freshwater and of this freshwater area, 10% is in the Canadian portion of the Great Lakes. Annually Canada's rivers discharge a volume of 105,000 m^3/s. Approximately 60% of Canada's freshwater drains to the north while 85% of the population lives along the southern border with the United States. Renewable water resources are not evenly distributed across the country and water shortages do occur during the growing season. Some parts of Canada are very dependent on groundwater resources (Fig. 4).

The principle water uses in Canada are thermal power generation (63%), manufacturing (16%), municipal (11%), agriculture (9%). Although agriculture is responsible for only about 9% of the total freshwater water withdrawals in Canada,

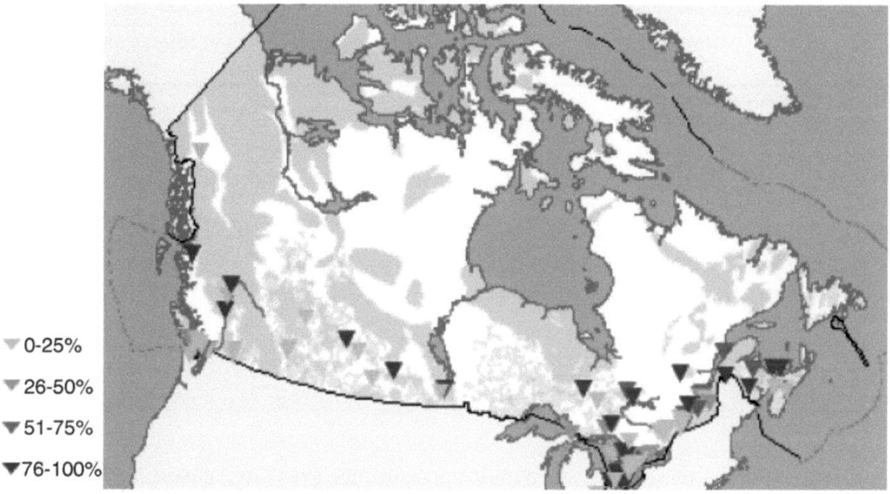

Fig. 4 Percentage of people using groundwater resources

it accounts for 70–80% of all water consumed. Irrigation accounts for the vast majority of the total agricultural water demand although livestock water use is significant in some regions.

Irrigation and Drainage

There are almost 1 million ha of irrigated agricultural land in Canada although a wide variation in climate and soil type results in an unequal distribution of this irrigated land among the provinces. Alberta, British Columbia, and Saskatchewan account for 85% of the irrigated land in Canada, as they have a large land base with fertile soils and a semiarid climate that makes irrigation essential for crop production (Table 2). In the province of Alberta, irrigation occupies 5% of agricultural land and contributes 18% to Alberta's agri-food GDP (AIPA 2002). In British Columbia, the agricultural land in the Okanagan Valley is 70% irrigated, accounts for about 8% of the total land in crops in B.C. but produces over 20% of the total value of the province's agriculture (BCMAF 2000; NRCAN 2006; Statistics Canada 2001). In Ontario, irrigation is required for the production of high value crops such as fruit crops and tobacco where diversification into specialty crops or crops that are drought sensitive results in higher value crops being produced (Shortt et al. 2006).

In Eastern Canada, with a wetter climate and an annual surplus of soil water, the issues of drainage, water quality and erosion control are of greater significance in terms of water management. Agricultural land drainage is needed to remove excess

Table 2 Estimates of potential and actual irrigated area in Canada

	Provincial estimates (ha)	Potential irrigated area (ha)	Potential irrigated area as % of actual
British Columbia	121,408	182,113	150
Alberta	660,777	800,000	121
Saskatchewan	80,939	404,694	500
Manitoba	30,352	60,704	200
Ontario	60,704	202,347	333
Quebec	25,000	35,000	140
New Brunswick	500	575	115
Nova Scotia	3642	7,285	200
Prince Edward Island	2023	4,047	200
Newfoundland	45	136	300
Total for Canada	985,390	1,696,901	172

soil water during the spring snowmelt and summer growing season and allow for crop production. There may also be periods of drought during the growing season in these regions where supplemental irrigation is beneficial, and sometimes essential, for the production of fruits and vegetables. Water table management or subirrigation which uses the subsurface drainage system to provide water to meet crop evapotranspiration requirements is often used. It has a water use efficiency in excess of 90% and is also very energy efficient. However, it is limited to very flat land slopes and soils with relatively good permeability. Some producers also collect the drainage water during periods of high drain flow and recycle this water into the subirrigation system to meet crop requirements.

Irrigation Improvements

Alberta and its 13 irrigation districts have invested CAN$ 850 million[1] since 1969 to upgrade 550,000 ha of irrigated land (Ring 2006). Water delivery became more effective as canals were repaired, lined and covered or replaced by pipelines, and canal automation evolved from local gate controllers to large SCADA (Supervisory Control and Data Acquisition) systems. On-farm water management improved as pump and sprinkler technology advanced, furrows were straightened and land levelled. In Alberta, there was an enormous increase in water application efficiency as farmers moved from surface to sprinkler to centre pivot irrigation systems. On-farm efficiencies increased from 36% in 1965 to 74% in 2000, as a result of adopting improved technology. It is also estimated that modernization and rehabilitation of infrastructure in Alberta between 1997 and 2000 resulted in a decline of return flows (as a percentage of gross diversion) of about 8% with a projected decline of 17% when all work is completed (Hart 2002).

[1] 1 CAN$ = US$0.944, average over 2008

Jurisdiction over Water Management

Different levels of government in Canada (federal, provincial, municipal) have different jurisdictional roles related to water management, while there are also many areas of shared commitment. The federal government has jurisdiction related to fisheries, navigation, federal lands, and international relations, including responsibilities related to the management of boundary waters shared with the United States and relations with the International Joint Commission. It also has significant responsibilities for agriculture, health and the environment, and ensures that national policies and standards are in place on environmental and health-related issues. Provinces have authority over waters that lie solely within their boundaries. Provincial legislative powers include flow regulation, authorization of water use development, water supply, pollution control and thermal and hydroelectric power development. Municipalities generally have authority over drinking water treatment and distribution and wastewater treatment operations of urban areas.

The Canada Water Act calls for joint consultation between the federal and provincial governments in matters relating to water resources. Joint projects involve the regulation, apportionment, monitoring or survey of water resources, and the pre-planning, planning or implementation of sustainable water resource programs. Between the provinces agreements are put in place as necessary. For example, a formal arrangement between the three Prairie Provinces (Alberta, Saskatchewan and Manitoba), called the Master Agreement on Apportionment, governs the sharing the waters of eastward flowing interprovincial streams and rivers in the Saskatchewan-Nelson watershed. According to the agreement, Alberta guarantees 50% of the natural water flow in the river system to downstream Saskatchewan, and Saskatchewan does the same for downstream Manitoba. The agreement also recognizes the problem of water quality and groundwater matters, and the Prairie Provinces Water Board administers the agreement and provides a forum to resolve and report on interprovincial water issues.

Canada and the United States share many waterways including the Great Lakes and rivers that mark or cross the border between the two countries. Decisions made within the water basins of one country can have consequences for the other. The International Joint Commission (IJC) sets the basic principles for guiding boundary water relations between Canada and the United States. The IJC helps anticipate, prevent, and resolve disputes between the two countries in an independent and impartial manner. It also provides a mechanism for cooperation and coordination in managing shared waterways and in investigating environmental issues of mutual interest along the border. This includes issuing Orders of Approval in response to applications for use, obstruction, or diversion of boundary waters; establishing boards for managing levels and flows of boundary and transboundary waters or for monitoring and assessing water quality in these waters; and carrying out investigations at the request of Canada and the United States to better understand an issue and to make recommendations to governments.

Conclusions

Managing water resources to supply water in sufficient amounts and of adequate quality to people, industry, agriculture and the environment is an ongoing and massive challenge in countries throughout the world. Increasing demand for water in the future due to rising populations, a greater need for food, richer diets that require more water to produce and environmental requirements will be further affected by climate change. Although conservation and more effective utilization of water resources can help to reduce future water scarcity, improving water storage, transboundary agreements between countries sharing the same water resources and interbasin transfer from water-rich to water-poor areas will make a significant contribution to providing sufficient, high quality and timely water supplies. Within Canada, significant improvements have been made to reducing the non-beneficial use of irrigation water and strict attention is paid to maintaining riverine environments. Water management agreements between Canadian provinces and with the United States have worked well in the past and will continue to be a very useful tool as demands rise for water use in North America.

References

AIPA. 2002. *South Saskatchewan River Basin. Irrigation in the 21st Century*. Report on line at http://www.aipa.org/21st_Century.html

BCMAF. 2000. *Irrigation Fact Sheet: Irrigated Area in British Columbia*. Order No. 550.100.1, British Columbia Ministry of Agriculture

FAOSTAT. 2004. Food Security Statistics. http://www.fao.org/faostat/foodsecurity/index_en.htm

Giordano, M. Zhongping Zhu, Ximing Cai, Shangqi Hong, Xuecheng Zhang and Yunpeng Xue. 2004. *Water Management in the Yellow River Basin:Background, Current Critical Issues and Future Research Needs*. Comprehensive Assessment Research Report 3. Columbo Sri Lanka, Comprehensive Assessment Secretariat.

Hart JR. 2002. *South Saskatchewan River Basin. Irrigation in the 21 century*. Alberta Irrigation Projects Association, Lethbridge, Alberta. 175 pp

IPPC. 2007. *Intergovernmental Panel on Climate Control Fourth Assessment Report*. http://www.ipcc.ch/ipccreports/ar4-syr.htm

NRCAN. 2006. *Okanagan Basin Waterscape*. Geoscape Canada. Geoscape.nrcan.gc.ca

Ring L. 2006. *Irrigation Infrastructure Rehabilitation in Alberta; 38 Years of Government/Industry Cooperation*. Presentation at the CSBE/SCGAB 2006 Annual Conference. The Canadian Society for Bioengineering, Edmonton, AB, Canada.

Shortt, R., Caldwell, W.J., Ball, J. and Agnew, P. 2006. *A participatory approach to water management: Irrigation Advisory Committees in South Ontario*. Can. Water Res. J. 31:13–24.

Stockholm International Water Institute/International Water Management Institute. 2008. *Saving water: from field to fork. Curbing losses and wastage in the food chain*. Stockholm International Water Institute/International Water Management Institute, Stockholm, 30 pp.

Statistics Canada. 2001. *Total area of farms, land tenure and land in crops, by province* (1981–2001 Census of Agriculture). www40.statscan.ca

UNESCAP. 2006. *United Nations Economic and Social Commission for Asia and the Pacific News Release*. March 2006. http://www.unescap.org/unis/press/2006/mar/g06.asp

United Nations. 2009a. *Population Prospects: The 2008 Revision*. Department of Economics and Social Affairs. Population Division. Available on-line at http://esa.un.org/unpd/wpp2008/peps_documents.htm

United Nations. 2009b. *World urbanization prospects: The 2009 Revision*. Department of Economics and Social Affairs. Population Division. Available on-line at http://esa.un.org/unpd/wup/index.htm

World Bank. 2006. Directions of Development. *Reengaging in agricultural water management. Challenges and Options*. World Bank, Washington, DC. 218 pp http://siteresources.worldbank.org/INTARD/Resources/DID_AWM.pdf

Integrated Water Resources Management

Vadim Sokolov

Origin of the Integrated Water Resources Management Approach

Throughout the development of civilization, people have come to the important realization that water, being the principal element of nature, calls for comprehensive management based on the integration of different waters, users, and impacts that determine the sustainability, efficiency, and safety of water availability. It is us, living on the cusp of a new millennium, whose rough lot it is to witness growing water deficits in nearly all parts of the Earth. Today, annual per capita fresh water resources available for use averages 750 m^3. By 2050, this value will decrease to an average of 450 m^3 even without taking into account climate change effects. This implies that over 80% of the countries worldwide will reach the UN water deficit level (Water: A Shared Responsibility. The UN World Water Development Report 2. UNESCO-WWAP 2006).

The fundamentals of the current integrated water resources management concept were put forward at the Dublin conference (1992) as four principles that served the basis for future global water reforms (The Dublin Principles for Water as Reflected in a Comparative Assessment of Institutional and Legal Arrangements for Integrated Water Resources Management. By Miguel Solanes and Fernando Gonzales-Villareal. TEC Background Paper No. 3 et al. 1999).

Principle 1. Fresh water is a finite and vulnerable resource, essential to sustain life, development, and environment. *Fresh water is a limited resource. This statement is supported by the quantitative analysis of the global hydrologic cycle, which suggests a fixed average annual volume of water. Fresh water is a natural resource that needs to be maintained by ensuring required water management. Water is needed for different purposes, functions, and services; therefore, water*

V. Sokolov (✉)
Scientific Information Center of Interstate Commission for Water Coordination (SIC ICWC), B.11, Karasu-4, Tashkent 700187, Republic of Uzbekistan
e-mail: vadim@icwc-aral.uz

C. Madramootoo and V.A. Dukhovny (eds.), *Water and Food Security in Central Asia*, NATO Science for Peace and Security Series C: Environmental Security,
DOI 10.1007/978-90-481-9974-7_3,

management should be integral (integrated) and take into account both demand for this resource and threats to its safety. This principle assigns a river basin or its catchment area as a water management unit, i.e. the so-called hydrographical approach to the institutional building of water management.

Principle 2. Water development and management should be based on a participatory approach that involves users, planners, and policy makers at all levels. *Water is a resource that affects all. True participation is ensured only when all stakeholders are drawn into the decision-making process. A participatory approach involving all stakeholders is the best way of achieving long-term accord and consensus. Participation means taking responsibility for an acknowledgement of the impact of sector actions upon other water users and water ecosystems, as well as assumption of liabilities to increase effective use and sustainable development of water resources. It should be noted that participation does not necessarily result in consensus; therefore, arbitration and other conflict resolution mechanisms are necessary too. Governments should work to create favourable conditions for the participation of all stakeholders, especially vulnerable groups of the population. It must be admitted that at present the poor will benefit least from mere establishment of a participatory environment unless the participation mechanisms are enhanced. Decentralization of decision making to the lowest level is the only strategy to enhance participation.*

Principle 3. Women play a central role in the provision, management, and conservation of water. *It is generally accepted that women play a key role in the collection and safeguarding of water for domestic purposes and, in many instances, agricultural use. At the same time, women play a less powerful role than men in the management, analysis of problems, and decision making related to water resources. IWRM demands that the role of women be acknowledged. In order to assure full and effective participation of women in all levels of decision making, it is necessary to consider the approaches that public agencies use to distribute social, economic, and cultural functions to men and women. There is an important relationship between gender equality and sustainable water management. Participation of men and women playing an influential role at all levels of water management can expedite the achievement of sustainability. Meanwhile, integrated and sustainable water resources management greatly contributes to gender equality by improving access of both women and men to water and water-related services, thus meeting their daily needs.*

Principle 4. Water has economic value in all its competing uses and should be recognized as an economic as well as social good. *In keeping with this principle, it is significant to recognize first the basic right of all human beings to have access to clean water and sanitation at a reasonable price. Managing water as an economic good is an important way to settle social tasks such as efficient and equitable water use as well as promotion of the economy and conservation of water resources.*

As soon as water is withdrawn from its source, it has a price as an economic and social good. Past failures to effectively manage water resources are associated with the failure to acknowledge the overall value of water.

Water Cost and Charge should be clearly differentiated. As a regulating or economic means, water *cost* in alternative uses is important to efficiently distribute water as a scarce resource. Water *charge* is used as an economic tool to support vulnerable groups, which influences them to save and use water efficiently, providing incentives to demand management, cost recovery of water services, and readiness of individual users to pay for extra water services. Recognizing water as an economic good is a significant decision-making tool to distribute water among different sectors of the economy and different users within each sector. This is especially important when water supply volume cannot be increased.

Goals of Integrated Water Resources Management

While declaring the goal of the IWRM principles as a milestone upon the attainment of which sustainable co-existence of the human being and environment is achieved, it should be born in mind that such integration should be ensured through the following three dimensions: area; vertical social hierarchy; and time (Duhovny and Sokolov 2005). It is a matter of fact that all stakeholders (beneficiaries) rather than water organizations alone should cooperate to achieve the integration of these elements and dimensions. Political circles and the backbone of the integrated sciences, such as natural, social, political, and technical sciences, should play the leading role.

IWRM is a new water management approach that promotes activity and collaboration of communities, society, and water users, and, at the same time, involves the government and local officials in achieving the Millennium Development Goals set forth at the UN Summit in 2000. Specifically, these are:

1. By getting equitable, sustainable, and guaranteed water availability, IWRM leads to reduced production losses that are due to disrupted water supply schedule and, consequently, is *conducive to an increase in income and reduction of poverty* (for example, in Malaysia, poverty has decreased from 60% to zero over the last 30 years).
2. By developing related sectors, attraction of internal and external investment, as well as water release, IWRM creates an environment for additional production, which *favours an increase in employment and income growth*. For example, the national income of the Kazaly and Aral regions of the Kzylorda oblast nearly doubled during 2002–2006 due to sustainable use of the Syrdarya River delta water.
3. By creating a sustainable drinking water supply and improving water quality in rivers and other sources, IWRM contributes to *better health of the population.*

4. By recognizing the principal role of water for ecosystems and ensuring water release for natural demands, IWRM promotes the recovery and protection of the environment.
5. IWRM enables coordinated use of water energy to increase hydropower production, which furthers a *stable power supply to the population during peak periods.*
6. *By involving many people, IWRM promotes water education and improves knowledge.*

What Is Integrated Water Resources Management?

The IWRM framework is based on a number of key principles that determine its practical essence (Duhovny and Sokolov 2005). These principles can be summarized as follows:

- Water resources are managed within hydrographical boundaries in accordance with the structure of a river basin.
- Water management includes inventory and use of all water resources such as surface, ground, and return waters, and takes into account regional climate.
- Close relation between all types of water consumption and stakeholders involved in water management horizontally across sectors and vertically across the water management hierarchy (basin, sub-basin, irrigation system, WUA, and farm).
- Public participation in the management, financing, maintenance, planning, and in the development of water infrastructure.
- Environmental needs are a priority in the activities of water authorities.
- Commitment of water organizations and water users to saving water and control of inefficient water losses; management of water demand along with control of resources.
- Information support; open and transparent water management systems.
- Economic and financial stability of water management.

It is important to expand upon the key principles in order to understand which steps should be taken to implement them.

Management Within Hydrographical Boundaries or According to a Hydrographical Indicator

It is well known that water knows no administrative boundaries. In accordance with physical laws, water goes through a hydrological cycle: it falls to the ground as precipitation and forms streams (rivers) from which it can be withdrawn for use, then evaporates into the atmosphere and turns into precipitation again. The area of landmass that hosts surface streams (rivers) is called a hydrographical catchment area (basin). Within the hydrographical basin, water continuously moves and naturally crosses different administrative boundaries that are manmade based on geopolitical considerations. Thus, it is clear that to control a variety of factors affecting the hydrological cycle, the entire river basin should be subject to management by a single water authority. If the organizational structure does not match hydrographical boundaries, this leads to mismanagement of certain elements of the hydrological cycle, which affects the stability of water supply and equitability of water distribution. The overall relationship of all levels of the water management hierarchy (Fig. 1) is based on the following two key conditions:

- Achieving potential productivity of water in all its uses at all levels of the hierarchy.
- Reducing unit costs of water in the system (as related to water intake) to the process requirement level (i.e. reduction of non-productive water losses).

Using All Types of Water Resources

Water resources used within a hydrographical basin are taken from surface and underground sources. The concern is that different authorities deal with the assessment of water formation in these sources; and most importantly, the use of that water is controlled by these authorities without required coordination. This can lead to information chaos when it comes to collecting data on the condition

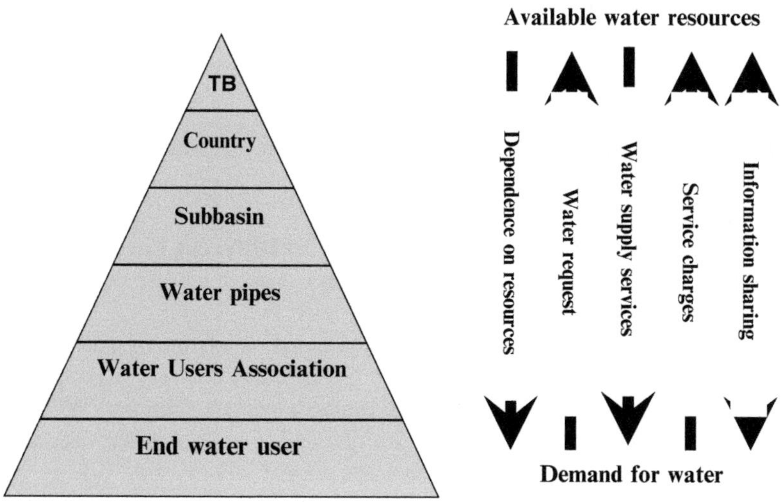

Fig. 1 Levels of the water management hierarchy and direction of the key interconnections between those in the IWRM system

of water resources, and results in somewhat anarchistic water use. As a result, non-productive water losses increase and water supply becomes uneven, while in some regions man-caused water deficits may take place.

Coordination of the Interests of Water Users from Different Sectors (Horizontally)

Under horizontal (intersectoral) coordination, water management authorities should fairly represent the interests of all water users from different sectors of the economy and ensure the priority of water saving and environmental preservation within each hydrographical unit. The main problem here is that different water resources are managed by different organizations. For example, surface water may be controlled by water authorities, advocating the interests of irrigated agriculture, and hydropower bodies, prioritizing the interests of energy producers. Geological institutions control the use of underground water. Public utilities and local administration are in charge of drinking water supply. Industrial water use is managed by relevant line ministries. There is no coordination between all the above mentioned government authorities.

The key condition for successful intersectoral integration is the conciliation of sector interests to enable joint use of shared water resources in accordance with agreed schedules, as well as one sector using the wastewater of other ones. When conflicting interests arise, conflict resolution mechanisms must be developed. This can be achieved through the participation of water users from different sectors in the public administration at appropriate levels of the water management hierarchy. The coordination tools are as follows:

- Joint planning and agreed use of water resources
- Coordinated development of the sectors
- Data exchange
- Sharing of the material and financial costs that are of mutual interest

Coordinating Water Resources Management at Different Levels of Water Management Hierarchy (Vertically)

As is known, the modern water management system is a scheme of multi-level water supply and distribution beginning from the basin, water delivery mains, secondary and tertiary canals, irrigation network of water users associations (WUA) or the water ducts of organizations of public utilities and industrial water users (Fig. 2). Most water losses, as well as troubles in water supply, are caused by disagreements between administrations at different levels of the hierarchy, which result in the overall inefficiency of the entire water management system. The problem is due to poor management quality rather than water deficit. Therefore, IWRM aims at

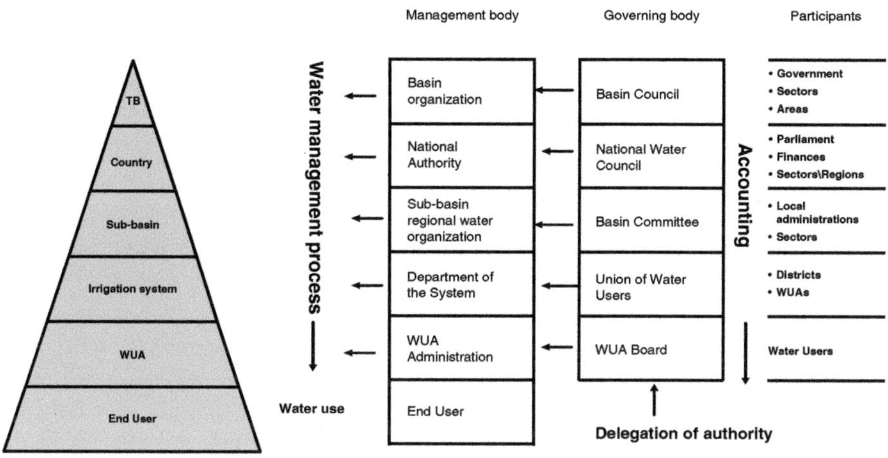

Fig. 2 Levels of the vertical water management hierarchy and key players

providing accord between different levels of the water management hierarchy. Moreover, it is necessary to eliminate the situation where each water management organization develops its own criteria and approaches that may disagree with the ultimate IWRM goal of getting maximum water efficiency.

Let us look at the modern water management structure (Fig. 2). The top level is represented by a basin and its sub-basins, then come systems of mains or managements of individual canals followed by the water user associations level (at land irrigation) or water user organizations (other users), and, finally, water users represented by farmers, businesses, residential areas, etc. If a river basin is within the country, then the river basin water organization (BWO) is set up as a part of the national authority that can include territorial offices for water resources management in the sub-basins.

Organizationally, these levels interact on the basis of requests for needed resources, completed "from bottom to top," and restrictions in the form of water supply limits and regimes formed "from top to bottom," these links are supported by agreements between BWOs and mains systems departments or between mains systems departments and WUAs (or public utility authorities), as shown in Fig. 1.

Public Participation in Water Resources Management and the Role of the Government

Wide involvement of NGOs and other stakeholders, such as local authorities, municipal water users, etc., in water management is a supremely important aspect of IWRM principles. Problems associated with water resources management must be discussed in the context of civil society-government relations. Public participation assures *transparency* and *openness,* thereby reducing the chances of making decisions that disagree with public interests. The more public participation, the less likely conditions arise that are conducive to corruption and neglect of the public interest. This is a tool to prevent parochial or departmental egoism in water use. It is also a platform for fair and crucial decisions related to water distribution under increasing water deficit conditions allowing for environmental conservation and development of society.

Ecosystem Approach: Environment as an Equal Partner in Water Use

People have long considered themselves 'all-powerful' and capable of mastering nature. However, the formula "We can await no gifts from the environment..." is now being replaced with the understanding that "We borrow nature from future generations rather than we are granted it from our ancestors." This approach has triggered the development of a global environmental movement and gradual development of environmental requirements, as well as environmental regulations to support sustainable links between people and natural complexes. In relation to the water industry, this means acknowledging rivers, lakes, and other water bodies as "water users" along with other entities. It also means that they lose their essence and purpose without certain water releases.

The environmental aspects of IWRM require understanding and action in two areas: preventing negative impacts of water and observing the requirements of nature and environmental complexes for water. The basic condition to ensure the sustainability of natural and natural-anthropogenic cycles is minimizing the quantity of negative impacts in the result of the interaction of water sources and economically exploited areas, as well as minimizing the interaction of surface and underground waters.

Kazakhstan Has Pioneered the Organization and Support of the IWRM Planning Process at a National Level

Looking back at the history of the initiation and mobilization of IWRM planning in Kazakhstan, one can state that the entire process took place according to a classical scheme (Fig. 3). At the turn of 2000, Kazakhstan had all the prerequisites for its water industry to transit to IWRM. At the same time, water management experts and decision makers came to understand that extensive institutional, legislative, and informational measures would have to be taken to ensure the development process and implementation of IWRM. Over 2000–2003, a supportive environment was created in Kazakhstan: in particular, the key role in water resources management was legislatively assigned to the Committee for Water Resources (CWR) of the Ministry of Agriculture and eight river basin water departments (BWD). However, considerable improvement in the institutional potential was needed for those organizations to operate efficiently. By 2003, a new water code was developed

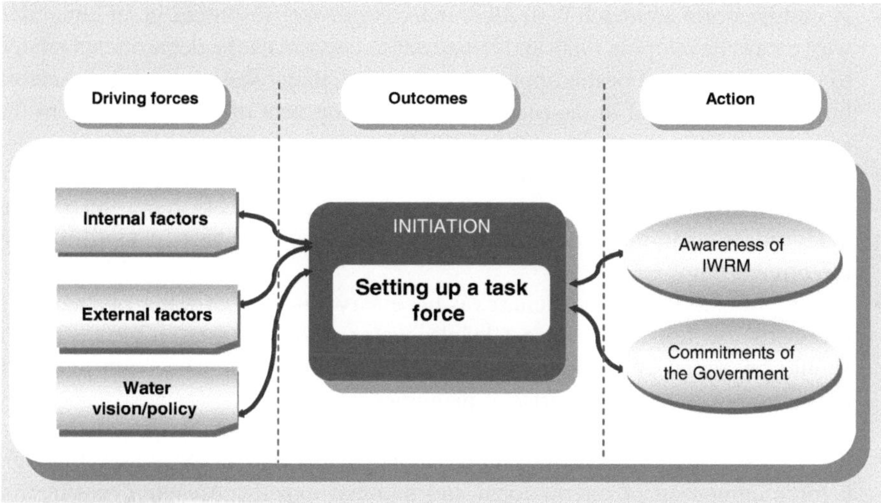

Fig. 3 Classical scheme of the initiation and mobilization of IWRM planning process (Integrated water resources management plans. Study guide and handbook. Published by CapNet & March of 2005)

and a package of direct-effect legal acts had to be developed to provide better influence on decisions made.

The IWRM planning process in Kazakhstan was initiated in a few areas. First, at the international level, Kazakhstan agreed to prepare a plan of the implementation of steady water resources management and development principles by 2005 at the World Summit for Sustainable Development held in Johannesburg.

This process was straightway supported by the international community and donors as represented by the Government of Norway, UNDP, and Global Water Partnership which were the driving forces for the preparation of the National Plan of IWRM in Kazakhstan. In the course of the official visit of the Prime Minister of the Kingdom of Norway, K. M. Bondevik, to the Republic of Kazakhstan on May 25–26, 2004, an agreement was concluded concerning financial support to Kazakhstan for the development of the National Plan of IWRM. Based on that, the Kazakhstan Prime Minister issued decree #302-p of October 13, 2004 as per which the Committee for Water Resources was charged to take the lead in the development of the National Plan of IRWM (NPIWRM) in association with other ministries and departments.

The development of the NPIWRM started in June 2004 with the assistance of the UNDP project "National Plan of IWRM and Water Efficiency in Kazakhstan," Norwegian Government, DFID (UK), and methodological support of the Global Water Partnership.

When initiating the development of the IWRM Plan, the project group realized that it would be different from a usual sectoral planning process. The main differences are as follows:

- A multisectoral approach is needed: managing water resources in an integrated way means developing links and structures to coordinate the departments related to water resources. For this approach to succeed, major water-consuming sectors have to get involved in the planning and development of the strategy from the very beginning.
- A dynamic process is required: creating a steady water resources management system and an integrated approach is a long-term process requiring regular analysis, adaptation, and, probably, modification of plans to ensure their effectiveness.
- Wide participation of beneficiaries is necessary: since most water management problems are felt at the lowest levels, water resources management changes should be aimed at the individual activities at the lower level with intensive consultations with beneficiaries at planning.

The first step in the course of the NPIWRM development project was setting up a task force composed of international and national experts. An interdepartmental task force (IDTF) was established to liaise with government structures. The group consisted of the representatives of the stakeholder ministries and departments invited by the Committee for Water Resources as follows:

- Committee for Water Resources of the Ministry of Agriculture
- Emergency Control Ministry
- Ministry of Economy and Budget Planning
- Ministry of Public Health
- Ministry of Environment
- Ministry of Energy and Mineral Resources
- Forestry and Hunting Committee of the Ministry of Agriculture
- Fishery Committee of the Ministry of Agriculture
- Department of Farming of the Ministry of Agriculture
- Committee for Rural Development of the Ministry of Agriculture

In order to raise public awareness and improve the knowledge of main project executors, a number of workshops, round tables, and training seminars were held in 2004–2005.

In March 2005, a concept of transition to integrated water resources management was developed and put forward to stakeholders. The concept provided an outline for NPIWRM and its principal components. The concept's development was guided by an international expert with experience in similar jobs in other countries and was sent to all relevant government agencies, institutes, local authorities, NGOs, and leading experts in Kazakhstan and other Central Asian countries to get comments with respect to the main elements of the future plan of IWRM.

The significant role that the project team under the UNDP office in Almaty played in implementing the action plan of the concept is noteworthy. The project team coordinated the work with the CWR and IDTF and discussions of any key issues that arose. The project team was completely responsible for ensuring public participation in the planning process and competently managed the activities required to prepare the IWRM plan. The project team has accurately performed the following objectives set:

- Set up and coordinate the overall process of the development of the national IWRM plan.
- Plan specific activities, workshops, training sessions, round tables, and meetings.
- Provide expertise and resources.
- Support the interdepartmental task force.
- Act as a coordinating center for communications.

The Kazakhstan experience has shown that the work of the team under direct guidance of a government body (First Deputy Chairperson of the CWR was the project director) is the best way to achieve long-term outcomes.

After having reviewed all comments on the concept, the first draft of the NPIWRM was prepared by November 2005 and submitted to all stakeholders for consideration. A significant success factor for any reform is political support and commitments at the highest level, i.e. the Cabinet of Ministers or Head of State. Perception of the need for sustainable management and development of water resources as well as the IWRM approach by political leaders is an important step in implementing the reforms. Sustainable management and development of water resources will significantly change traditional water management methods, while the necessary alterations in institutional and decision-making systems will have certain political effects. This should be recognized and taken into consideration when planning IWRM. Real political support is necessary because this ensures:

- Priority water management problems can be settled at the interdepartmental level.
- Effective planning coordination as the interdepartmental task force will rely on political support in its work.
- The water resources development vision takes into account political goals compatible with other national development goals.
- And vice versa, water resources management and objectives are taken into consideration in the political agenda.
- Sustainable water management approaches are included in the national development plans, activities, and political statements in other sectors.
- Political effects of the IWRM plan are allowed for throughout the entire process rather than at a formal end stage (thus ensuring ongoing improvement of the works).
- Decisions are made according to the suggested plans as well as legislative and institutional reforms.
- The IWRM plan will be adopted and implemented.
- Allocation of government funds, and, if needed, mobilization of donor assistance.

Based on the abovementioned considerations, during the initial review of the first draft NPIWRM, the IWRM IDTF members recommended adjusting the plan in accordance with the Regulations for the development and implementation of sectoral and regional programs approved by the Decree # 231 dated February 26, 2004 of the Government of RK.

The Decree # 978 dated October 11, 2006 of the Government of RK "On signing agreement between the Government of the Republic of Kazakhstan and UNDP concerning the project "National Plan of Integrated Water Resources Management and Water Efficiency for the Republic of Kazakhstan" approved the development of the Program "Integrated Water Resources Management and Improvement of Water Use Efficiency in Kazakhstan till 2025." The Program is divided into two phases. The first stage (years 2008–2010) provides for:

(a) Settling the following tasks related to water resource management improvement:

- Assistance with the development and implementation of the national water policy of Kazakhstan.
- Increasing the organizational capacity of CWR and BWDs and their authority areas.
- Setting up a network of basin councils and extending their power.
- Working out basin water resource management programs for each river basin in Kazakhstan.
- Development of basin agreements and mechanisms of their observance.
- Introduction of scientifically substantiated standards to limit water use and wastewater discharge.
- Development and introduction of mechanisms of damage recovery for pollution and lack of coordination at water facilities.
- Introduction of a charging system for ecosystem-based services.
- Development of a monitoring system for water bodies and land resources.
- Establishing a framework for the introduction of a unified information and analysis system, advancement of access to and sharing of the information related to water management and conservation.
- Informing and ensuring participation of all relevant population strata in addressing water issues.
- Professional development and training of water sector specialists; establishment of training centers.
- Strengthening intergovernmental cooperation in the use and preservation of transboundary water bodies; conclusion of agreements on information interchange and establishment of intergovernmental basin databases for complex use and conservation of transboundary water resources.

(b) Decide on the problems related to water use efficiency improvement:

- Creating an institutional and legal environment to set up associations of water users, irrigation and drainage condominiums, and water efficiency advisory services.
- Introduction of economic incentives and technological innovations.
- Development of an action plan to improve the monitoring system of water bodies and water facilities at the local level.
- Conducting research to identify effective irrigated agricultural methods and training farmers.

- Development and implementation of information programs/campaigns on water resource management and water saving.

The second phase (years 2011–2025) aims at:

(a) Meeting the following challenges related to the improvement of the water resources management system:

- Raising the institutional capacity of the organizations working on water resources management and conservation.
- Developing intersectoral cooperation mechanisms.
- Ensuring water management planning at inter-governmental, national, basin, and territorial levels.
- Developing and introducing economic incentives for sustainable water use.
- Improving an ecological standardization system as a part of the environmental security system.
- Developing a water bodies/ecosystems and natural water quality monitoring system.
- Creating and developing information infrastructure for water resources management and conservation.
- Advancing the education system and advanced training of specialists in water resources management and conservation.
- Wider international cooperation and enhancement of transboundary water management.
- Implementing the Agreements on information interchange and establishment of basin and international databases for complex use and conservation of transboundary water bodies.

(b) Solve the following problems related to water use efficiency improvement:

- Development of a network of water user associations at the local level (WUA, UCWU) and water efficiency advisory services.
- Upgrade charge-based water use mechanisms taking into consideration the establishment of associations of water users and water transportation services rendered by them.
- Improve the water bodies and water management monitoring systems at the local level.
- Develop and implement pilot projects on water use efficiency improvement.
- Conduct information campaigns and training workshops on water saving.

Special attention should be paid to the institutional measures at all management levels, which are planned within the project. According to the current legislation of the Republic of Kazakhstan (Article 33, Water Code), water resources inventory use and conservation management is carried out by the Government of Kazakhstan, (the agency authorized for the water resources inventory use and conservation), local representative and oblast (regional) executive bodies (cities of national status, capital cities) within their jurisdiction fixed by the laws of the Republic of Kazakhstan. Based on the integration and hierarchy, water resources inventory use and conservation management is structured into the following levels: (1) intergovernmental; (2) national; (3) basin; and (4) territorial.

Kazakhstan's water sector is managed based on the watershed principle. The key responsibility for public management is borne by the authorized water resources inventory use and conservation agency (CWR) and its regional basin departments (BWD). The Program aims at the gradual improvement of the organizational basin management structure. At the initial stage of the Program, as shown above, it is suggested a draft resolution be submitted for consideration by the Kazakhstan Government concerning the promotion of CWR and RBA status to carry out strategic functions securing the development and effective realization of water resources inventory use and conservation policy. Inadequate status of CWR as an authorized water resources inventory use and conservation agency impedes the equitable negotiations of CWR management with ministers, deputy ministries, and even with akims (heads of local authorities). The only way to ensure effective control of water use and efficient water resources inventory management is to have the power to "dictate water policy." At the national level, water users are represented through different ministries. Therefore, the status of the CWR should be promoted to at least the level of other ministries related to water resources use and conservation. A similar situation is observed at the basin level. Lack of adequate status prevents BWDs from effective performance of its functions as per the Water Code. Establishment of Basin councils worsens the situation even more, as the chairperson of the Basin council is, at the same time, Head of the BWD and has a status lower than oblast akims.

In general, the following measures should be taken to improve the institutional environment:

In the initial phase (years 2008–2010)

- Submit a draft resolution of raising the status of the CWR and BWD for consideration by the Government.
- Include the first deputy chairperson of the CWR in the Sustainable Development Council (SDC).
- Optimize the organizational structure of the CWR by setting up divisions on the implementation of the IWRM Program, raising public awareness, handling legal and environmental issues; establishing a control service for operational, annual, and long-term water management planning, as well as analysis and adjustment of water use plans.
- Improve the structure of water management bodies at the basin level (BWD).
- Develop and introduce a scheme of BWD actions coordination with other organizations and departments related to water management.
- Delineation of the responsibilities between authorized agencies for water resources inventory use and conservation, agencies for environmental management, mineral resources investigation and use, local executive agencies, etc.

The second phase (years 2011–2025) provides for:

- Heightening the status of the CWR and BWD; based on the discussion concerning the options to raise CWR and BWD statuses under the project, the best option found was to reorganize the CWR into the Ministry of Water Resources to ensure direct decision making in water resources management. This will allow integrated water resources management to be carried out and mobilize management functions and tools to tackle water sector problems at the intergovernmental, national, basin, and local levels.
- Support and develop the basin councils network.
- Delegation of the underground fresh water management functions from the Ministry of Energy & Mineral Resources to the Central Water Management Executive Agency of the Ministry of Water Resources.
- Assignment of the hydraulic facilities management functions to the Ministry of Water Resources.
- Founding a committee for public water supply and water disposal under the Ministry of Water Resources. The Committee's activity should aim at the improvement the tariff policy and the legislative framework for water supply and water disposal, identification of funding (subsidization) sources and supporting water supply and disposal plants in solving immediate tasks and long-term ones, including an issue related to the inventory of water and sewage system, monitoring of public utility facilities, introduction of new technologies, personnel training, etc.

In order to develop the organizational capacity, the project recommends the following measures.

At the first stage (years 2008–2010):

- Increase the staff size of the water resources inventory use and conservation agency (Committee for Water Resources).
- Increase the staff size of the river basin water departments and provide with adequate resources.
- Search for potential funds for the national water management system.
- Found a water users support and consulting department under the Research Center of Water Resources Sector; deliver training in water-saving technology application; and promote environmental awareness among the population.
- Improve the organizational structures of all institutions involved in water resources management.

At the second stage (years 2011–2025):

- Review the current financial allocations and their potentials.
- Distribute finances for the water resources management sector.
- Further improve the CWR and BWD structures and increase the staff, providing them with adequate resources.
- Develop the organizational capacity of other organizations involved in water resources management.

Summary

Kazakhstan's IWRM is an excellent example of how such programs can be adopted by all beneficiaries, including the government, as a long-term national program. There is no point in wasting resources on the development of a plan that will be rejected or "shelved," and never implemented. For that very reason, political support and involvement of beneficiaries from the very beginning of the IWRM planning are vital. If the participation process is well-organized, then approval will easily follow, because at the first meetings the project executors will have come to agreements with beneficiaries concerning all the conditions that must be provided. From the outset, the project team monitors how much the process suits the conditions required the plan to be approval.

Reaching a consensus regarding the development of the plan and its content increases the chances that the plan will be approved by beneficiaries and the government. If all beneficiaries, including the government, are involved in the development of the plan right from the start, approval is going to be just a formality.

Maintaining contacts with beneficiaries and the government throughout the development of the plan should be a part of the strategy developed by the task force for the entire planning process. The final version of the IWRM plan should be made public and easily accessible. This is significant, since it is impossible to work with all beneficiaries without consultation. Most consultations can be selective. As soon as the national plan is adopted, it is important that everyone can access it, dispute it, and get ready for its effects. Today, all project materials are available on a special web page www.voda.kz

It should be noted that the interdepartmental task force of the project did not recommend adoption of any particular government document concerning the implementation of the IWRM program at the end of the project. In his speech at the Republican Water Conference in Almaty on October 26, 2007, CWR Chairman A.D. Ryabtsev spoke about the decision made on the necessity to integrate respective issues into various programs and plans of stakeholder ministries and departments being implemented or developed.

References

Duhovny V.A., Sokolov, V.I. Integrated water resources management. Central Asian experience and lessons toward the Fourth World Water Forum. Tashkent, 2005.

Integrated water resources management plans. Study guide and handbook. Published by CapNet, March of 2005 (translated into Russian), 104 p.

The Dublin Principles for Water as Reflected in a Comparative Assessment of Institutional and Legal Arrangements for Integrated Water Resources Management. By Miguel Solanes and Fernando Gonzales-Villareal. TEC Background Paper No. 3, Global Water Partnership, Stockholm, Sweden, 1999.

Water: A Shared Responsibility. The UN World Water Development Report 2. UNESCO-WWAP, 2006.

Food Security and the Improvement of Water Use Efficiency

Said Yokubzod

Population growth and an increasing demand for water resources connected with the development of the world economy and extensive use of land and water resources in developing countries are gradually leading humankind to global water scarcity and a food crisis. This situation is complicated because people have begun to use agricultural products, particularly grain, for producing energy due to fuel shortages and the rise in oil prices. Under market conditions it is very difficult to stop the use of food products for fuel through administrative measures.

Central Asia belongs to a region with a high rate of population growth. According to World Bank data, the total population of countries within the Aral Sea Basin will be 173 million by 2050 (Table 1).

At present, there is no food crisis in the Central Asian countries of the Commonwealth of Independent States but a shortage of food is experienced all the time. Given such a situation and the lack of strategic reserves, any destabilizing factor can upset the fragile balance of food in the region. For example, in Tajikistan early snowfall and a lengthy frost during the past winter did not allow wheat to be planted and destroyed winter potatoes, cold-sensitive fruits and vines. The Government of Tajikistan, anticipating the coming food crisis, has requested international organizations about extraordinary help.

Theoretically, no state is capable of ensuring full and sustainable food security for their population over the long-term since food demand depends on supply. However, countries with developed agricultural production not only provide their people with food but are also the main suppliers of food products all over the world.

There are a number of factors affecting the food self-sufficiency of a country:

- Effective agricultural production development policy;
- Effective cropping patterns aimed at supplying people with food;
- Regular improvement of water and land productivity by applying intensive technology to agricultural production;

S. Yokubzod (✉)
Minister of Land Reclamation and Water Resources of the Republic of Tajikistan,
5/1 Shansi Str., Dushanbe, 734064, Republic of Tajikistan
e-mail: taj_water@mail.ru

C. Madramootoo and V.A. Dukhovny (eds.), *Water and Food Security in Central Asia*,
NATO Science for Peace and Security Series C: Environmental Security,
DOI 10.1007/978-90-481-9974-7_4,

Table 1 Expected population growth in Central Asia

Central Asian States	Population[a], million 2006	2050	Expected increase, % 2050
Afghanistan[b]	24.5	97.3	297
Kazakhstan[b]	14.8	13.1	−11.5
Kyrgyzstan[b]	5.3	6.7	26.4
Tajikistan	7.0	10.4	57.6
Turkmenistan	4.9	6.8	38.8
Uzbekistan	27.0	38.7	43.3
Total (excluding Afghanistan)	59.0	75.7	28.3
Grand total	83.5	173.0	108

[a]Data from UNPFA
[b]The whole territory, including outside the Aral Sea Basin

Table 2 Total and unit irrigated area per capita in Central Asian countries

Central Asian countries 2000	Area under irrigation (thousand hectares)	Unit area (ha/capita)
Kazakhstan	786	0.30
Kyrgyzstan	422	0.14
Tajikistan	740	0.11
Turkmenistan	1,735	0.41
Uzbekistan	4,295	0.19
Total	7,971	

- Development of a framework for tax and legal incentives encouraging the use of highly effective and water saving technologies for growing crops;
- Regular expansion of irrigated land area according to population growth;
- Other factors.

In Central Asia, where outdated irrigation technology is used, food security depends on irrigation. In our region (excluding Afghanistan) the total irrigated area is about 8 million ha (Table 2). In order to develop new land, it is necessary to have available water resources that we do not have. More than 90% of the water resources of the Amudarya and Syrdarya Rivers in the region are used for irrigation. Under the current level of irrigation technology we now experience shortages of water and moreover water supply to fields decreases sharply in dry years.

Whereas in the 1980s the cost to put new land under irrigation was US $3,000–8,000 per ha, now it is already up to $12,000–15,000 per ha. Given such a tendency, by 2020, the cost of one hectare of irrigated land could be more than US $ 20,000. However, there is another major problem that will restrain the development of new lands. It is lack of available water resources. Even now, countries in the flow dispersion zone have depleted their prescribed water withdrawal limits from the Amudarya and Syrdarya. Evidently, withdrawals of water exceed their limits in some places. Otherwise it is difficult to explain why the yearly formed

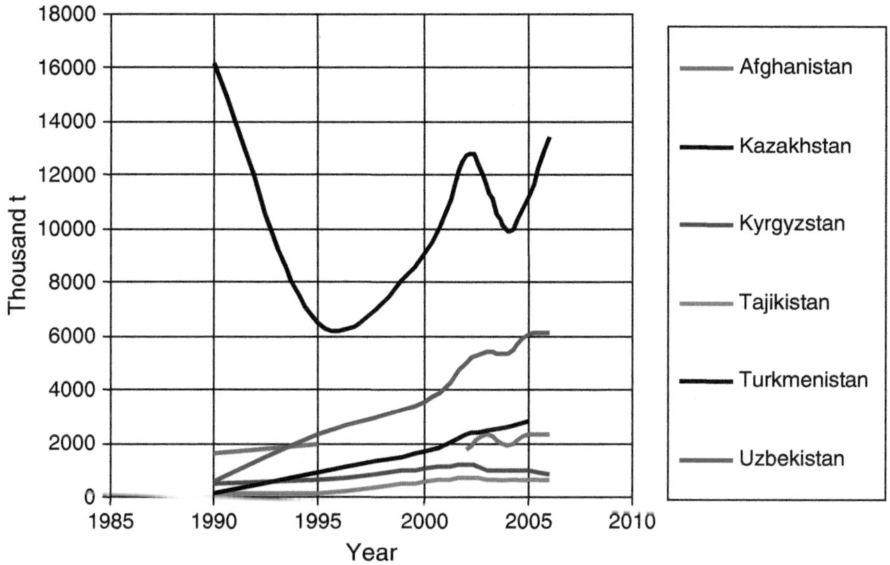

Fig. 1 Wheat production in the Central Asian countries

plans of water supply to the Aral Sea from the Amudarya River are not frequently implemented. It is also necessary to emphasize that water delivered to the Aral Sea is not very fresh.

Production of major agricultural products to ensure minimal food security, i.e. of wheat, in the region's countries strongly depends on available croplands (Fig. 1). In this context, there are different conditions across the region: Kazakhstan has a huge land area where wheat grows on dry land and is the main exporter of wheat in the region; Uzbekistan and Turkmenistan possess enough irrigated area that they could produce sufficient amounts of cereals for their people for a short period. Difficulties with production of sufficient cereals remain in Kyrgyzstan and Tajikistan which have the lowest irrigated area per capita in the region. Tajikistan annually imports up to 700,000 tons of cereals.

One of the ways to attain food security for the region's vulnerable countries is to develop new lands and increase agricultural production. For example, there are up to 800,000 ha of land suitable for irrigation. Simple analysis shows that to achieve the average regional value regarding unit irrigation area of about 0.2 ha per capita, Tajikistan has to develop 650,000–850,000 ha of land by 2015. To this end, it is necessary to put into rotation 10,000 ha of new irrigation lands. In the near future, we have no such aim.

However another economically more beneficial and ecologically friendly solution of the food security problem, is to increase the productivity of each irrigated hectare and of water. Increasing water productivity is a twofold task: increase soil fertility and crop yields and conserve water. This is a complex problem and we are just starting to solve it in Tajikistan.

Observations during 2002–2004 in North Tajikistan within the framework of the project "Integrated Water Resources Management in Fergana Valley" funded by the Swiss Agency for Cooperation and Development highlighted that we still have low water productivity. According to data from SIC ICWC, over the last 25 years water productivity has decreased two to three times (Fig. 2).

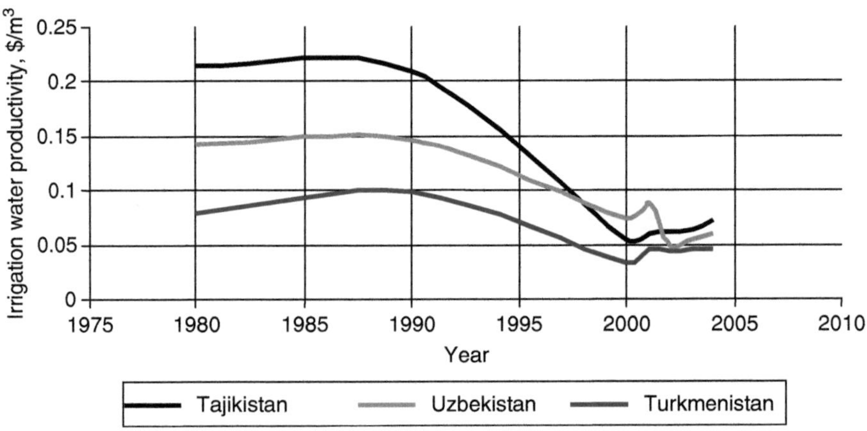

Fig. 2 Irrigation water productivity in Tajikistan, Turkmenistan and Uzbekistan

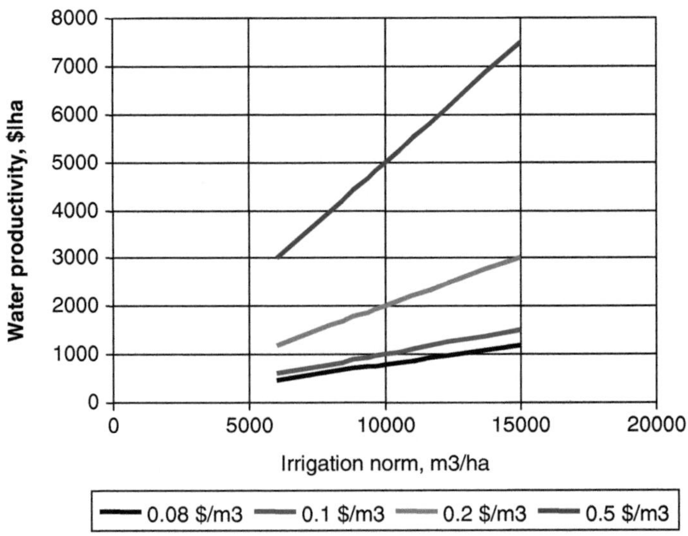

Fig. 3 Potential increase of water productivity under various levels of irrigation rates in Central Asian countries

World experience shows that only improvement of water and land efficiency can be a guarantor for food security.

Water productivity is a universal indicator, and increasing it can provide sustainable development of agricultural production, robust performance of irrigation and drainage systems, and finally, market saturation and food security (Fig. 3). In order to increase water productivity, we have to develop a long-term program for increasing the effectiveness of water and land use, including the solution of the following problems:

In Water Resources Management

1. Completion of the transition to the basin principle of management, establishment of basin governance for management of canals and involvement of all water users in water management.
2. Reconstruction, rehabilitation and modernization of irrigation and drainage systems at the inter- and on-farm levels in order to improve system and management efficiency.
3. Finalization of the process of establishing Water User Associations on the basis of private farms and at on-farm level.
4. Improvement of framework and approaches in water resources management at all levels.
5. Application of differentiated charges for water services; improvement of economic mechanisms for water resources management.
6. Creation of suitable conditions for organization of production and implementation of water saving technology allowing for application of nutrients together with irrigation water and for significant increase of crop yields.
7. Implementation of improved irrigation technologies, including application of conventional furrow irrigation allowing for decreased surface and drainage outflows on the fields.
8. Restoration of a professional training system for hydraulic engineers and training for irrigators.
9. Development and implementation of a system of material and moral incentives for water saving at all hierarchical levels of water management.
10. The State Long-term Program for Water Sector Development coordinated with agricultural production development.

In Agricultural Production

1. Development of an economic model and legislative support for its adoption for the balanced and stable turn-over of capital in irrigated agriculture, taking into account as far as possible, all factors listed above.

2. Development of an enforcement system for obligatory allocation of a certain part of farm income for O&M of irrigation and drainage systems.
3. Government economic support for farms situated in the pumping irrigation area; granting of soft loans to them for application of water saving technologies.
4. Revision of cropping patterns aimed at high-yielding crops, subject to local and international markets, especially in the pumping irrigation area.
5. Establishment of agroservice and consulting centers (extension services) to support farmers with:
   (a) Application of highly efficient agricultural machinery
   (b) Marketing of high-quality seeds, fertilizers, and sale of produced products
   (c) Research and development of recommendations for the improvement of soil fertility
   (d) Development of measures to control agricultural pests and apply pesticides safely
   (e) Application of water saving and land leveling technologies
   (f) Organization of training for farmers

Fulfillment of the above mentioned tasks will create a legislative, economic and institutional basis for the improvement of water productivity and for ensuring food security. However, it is easy to note that all these measures are implemented in countries with advanced irrigated agriculture such as USA, European countries, Israel, etc.

Achieving food security is also promoted by regional differentiation of agricultural production. Therefore there is a need for close and stable long-term economic integration of the countries in the region, based on trust for one another. For example, on the basis of a regional Food Agreement, Kazakhstan could supply needy countries with wheat, milk and meat, and Tajikistan with fruits, vegetables, water and electricity.

The President of the Republic of Tajikistan Emomaly Rakhmonov said that only the completion of the Roghun hydropower plant with a capacity of 3600 MW and a reservoir of more than 13 km^3 will allow the protection of more than 3 million ha of land from severe water shortage in the lower Amudarya. Given global warming, dry years will occur more frequently and floods will bring not only damage but loss of water – a main resource of food production. The drought during 2000–2001 demonstrated to us the danger of such phenomena. Using words by our President we are sure that a water partnership, despite many difficulties in the Central Asian region, needs to be protected and strengthened comprehensively. It is necessary to use the potential of water widely for the benefit of creation.

Security and Water Resources Management Problems and Experience in the Amudarya River Basin

Yuldash Khudayberganov

Water Resources Management

Introduction

Until 1987, the USSR Ministry for Agriculture and Water Resources (MAWR), Republican MAWRs and Upradik (a water management organization under USSR MAWR, responsible for Republican water sharing in downstream rivers based on principles of rigid objectivity, and located in Urgench) played a key role in water resources management and regulation in the Amudarya basin.

A need for integrated water resources management and protection at the basin level was rationalized long before the independence of the countries in the region. Although USSR MAWR implemented a centralized water sharing system based on consultations with five governments, analysis of the consequences of water shortage in 1974–1975 and especially in 1982 showed that strictly controlled environmental and quantitative water supply was impossible without a single water management organization for the whole basin. Thus, in October 1987, the Basin Water Organization (BWO) "Amudarya" with headquarters in Urgench was established by USSR MAWR Decree upon recommendation of the Republics in the region to improve and increase the effectiveness of water resources management in the basin, move to basin-wide management, as well as assure more operative, well-timed solutions of water-related problems. BWO is responsible for operative water resources management and regulation among the Republics, well-timed and regular water supply to users within fixed limits (agreed to by governments), sanitary and ecological releases to the Priaralie area and the Aral Sea.

Interstate canals and all head waterworks on rivers and main tributaries with a discharge greater than 10 m^3/s were transferred to BWO's responsibility, according

Y. Khudayberganov (✉)
Basin Water Organization Amudarya,
63 Al-Zamakhshary Str., 140000, Urgench, Republic of Uzbekistan
e-mail: amu_bvo@yahoo.com

C. Madramootoo et al. (eds.), *Water and Food Security in Central Asia*,
NATO Science for Peace and Security Series C: Environmental Security,
DOI 10.1007/978-90-481-9974-7_5,

to the Government Decree 1110. BWO also controls some head works not under its responsibility, all pumping stations taking water from rivers and main interstate canals, and reservoirs, main river gauging stations, and return water discharge into the Amudarya River. According to the general agreement with the countries in the region, BWO "Amudarya" is responsible for managing and controlling water diversion structures located in the following transboundary river basins:

1. Pandj River
2. Vakhsh River
3. Kafirnigan River
4. Amudarya River

Under present conditions, the Amudarya River flow is poorly regulated. This leads to river water stress and complicates the use of this water for economic purposes. Active Nurek and Tyuyamuyun reservoirs are regulated seasonally. At present, they are the main regulators of the Amudarya River flow. In addition there are a number of in-basin and in-system reservoirs distributing and redistributing river flow in the Amudarya River basin. At present, water is diverted from the river by means of dams or without them and some water is taken by lifting.

Energy, the fishing industry and water transport are the key water users in the Amudarya River basin. Industry, municipalities and agricultural irrigation are the major water consumers. The main agricultural crops are cotton and grain. Gardening, wine-growing, silkworm breeding, beet and melon growing are also developed in the Amudarya River basin, while rice growing is promoted in the river downstream. The large-scale desert pastures are used as a forage base for sheep breeding, mainly for caraculs. Irrigation is the largest element of the water management system.

Taking into account morphological and geographical characteristics, the Amudarya River basin is divided into three sections: upstream (upstream of Kelif – border between Turkmenistan and Uzbekistan), midstream (between Tyuyamuyun and Kelif) and downstream (downstream of Tyuyamuyun). The total irrigated area is 4.0–4.5 million ha. Upstream is comprised of the irrigated lands of Tajikistan, Uzbekistan (Surkhandarya province) and Kyrgyzstan (small irrigated area in the south of the Republic). Thus, upstream irrigated lands are located in valleys of the main components of the Amudarya and its tributaries: Pandj, Vakhsh, Kafirnigan, Surkhandarya and Sherabad. Today, the largest midstream irrigated areas are focused along extended canals such as the Garagum canal, Karshi main canal with a cascade of 6 pumping stations and the Amubukhar canal. Off-stream reservoirs function in each canal system. Irrigation systems from Kelif up to Tyuyamuyun take water from tens of canals, diverting water without a dam. Large-scale canal systems such as Tashsaka, Pakhta-arna, Klychniyazbai, Urgench- Daryalyk-arna, Kipchakbozsu, Khan-yab (Sovet-yab), Dzhumabaisaka, Kyzketken and Suenli were built downstream in the Amudarya River along two river banks. Canal systems such as Tashsaka, Klychniyazbay and Kipchakbozsu are of interstate importance.

The existing water management system in the Amudarya basin is a complex of structures assuring water resource transportation, flow transformation, water withdrawal and water supply to water users, power generation, and the control, accounting and quality of water resources. The water management system largely defines conditions of economic development in a given region and consists of natural and economic objects that form, transport, redistribute and consume water resources. Complexity of the water management system is caused by the fact that management infrastructure is dispersed within four sovereign Central Asian states. Specifics of the water management system can be formulated as follows:

- A huge volume of heterogeneous information characterizing conditions of the water management system
- Abundance and spatial dispersion of management bodies and information sources
- Stochastic nature of hydrological information
- Inconsistent demands of the system's actors regarding management regimes
- Lack of common economic criteria for water resources use

First of all, interstate (transboundary) river waters of the basin are included in management.

It should be underlined that the responsibility for large-scale head intakes (36 units) were allocated to the BWO Amudarya, excluding gauging stations such as the Dangarin (Tajikistan) and Garagumdarya (Turkmenistan), and the Tyuyamuyun waterworks facility (LBK, Turkmendarya, PBK, Pitnyak-Arna) – Uzbekistan, which were to be transferred to BWO after completion of their construction. Four administrations for the operation of intake structures, waterworks facilities and interstate canals were established under BWO Amudarya with centers in Kurgan-Tyube city (Tajikistan), Turkmenabad city (Turkmenistan), Urgench city (Uzbekistan) and Takhiatash city (Republic of Karakalpakistan) so that BWO could manage transboundary water resources.

Territorial coverage is divided in the following way between the administrations:

- Upstream Administration operates intake structures and controls water withdrawals from Vakhsh, Pandj and Kafirnigan and within the Amudarya River reach of 246 km, up to the gauging station Kelif.
- Midstream Administration controls water withdrawals within the Amudarya River reach of 552 km, between gauging stations Kelif and Darganata
- Administration for Amudarya inter-republican canals (Upradik) operates river water intakes, maintains and operates interstate canals with their structures, controls water withdrawals within a river reach from the Tyuyamuyun hydro unit up to the gauging station Kipchak (reach is 167 km). Upradik manages three large-scale irrigation systems:

    1. Tashsaka
    2. Klychniyazbai
    3. Kipchak-Bozsu

- Downstream Administration operates Takhiatash waterworks facility, head intake structures on Khan-yab and Dzhumabaisak and controls all water withdrawals within reach from the gauging station Kipchak up to the Aral Sea (reach is 283 km).

According to the Statute, BWO should perform the following functions in order to achieve set goals and objectives:

- Develop water withdrawal plans, operating mode for cascade of reservoirs, correct and agree upon seasonal water consumption limits for riparian states by ICWC.
- Medium-term planning, joint water resources development and protection that is agreed with the water management and power departments of riparian states, and participation in perspective planning.
- Supply water to the state: water consumers, the Aral Sea and Priaralie, according to ICWC decisions.
- Provide monthly information on water resource use to ICWC members.
- Develop and implement an automated water resources management system in the Amudarya River basin, water accounting and water measuring within head intakes, providing them with automated and telemechanical equipment.
- Jointly with hydrometeorological services, perform control water measurements within boundary sites of territorial administrations in order to make a balancing account of river flow.
- Control observation of operating mode of reservoirs' cascade as agreed by ICWC.
- Maintain, reconstruct and operate waterworks facilities, head intakes, interstate canals and collectors, infrastructure of automated water resources management system in the Amudarya River basin and other infrastructure which are under its responsibility.
- Jointly with water management bodies and other interested enterprises and organizations, develop and implement measures to control floods and protect settlements and agricultural lands from water-logging, flooding and other water-related disasters.

BWO "Amudarya" is guided by a BWO Statute agreed to by ICWC and acting legislations of ICWC member countries as well as ICWC decisions, agreements, protocols and other regulations. BWO is financed at the expense of allocations of three states (Uzbekistan – in soums, Tajikistan – in somoni and Turkmenistan – in manats).

Operation Pattern

Water use is planned in the following way under the Amudarya water management system. BWO prepares proposals for water withdrawal limits according to a water availability forecast for the planned period provided by the Uzbekistan

Hydrometeorological Service. Water withdrawal limits for each ICWC member-state for a hydrological year (nongrowing and growing seasons) are discussed and agreed upon during ICWC meetings. BWO prepares proposals for water sharing limits. After approval of limits, BWO collects water withdrawal requests for transboundary river flows and develops operating modes for the cascade of reservoirs (Nurek and Tyuyamuyun). An operating mode for the Nurek reservoir is prepared jointly with UDC (United Dispatcher Center) "Energy" and it should be emphasized that UDC has priority when preparing the operating mode. The operating mode for the cascade of reservoirs is discussed and agreed upon during ICWC meetings. Operating modes for reservoirs are developed for a planned period. Then, correspondence between actual water withdrawals and approved limits and requests of each ICWC member-state is checked. Line services provide information on actual water withdrawals to the Control office of BWO territorial administration. Information is submitted to BWO's central control centers after checking information reliability.

BWO "Amudarya" prepares materials related to the establishment of water withdrawal limits for the Amudarya River for each country and water consumer for discussion at ICWC meetings:

- A country submits its proposal for decade water withdrawal limits by province for a particular period to BWO "Amudarya" 10–15 days before the beginning of a respective period.
- BWO "Amudarya" agrees to water withdrawal limits for each state – consumer on-site jointly with provincial water management organizations based on applications provided by countries and within allocated limits.
- Generalized materials on water withdrawals in each country are reviewed and agreed to at regular ICWC meetings.
- BWO "Amudarya" revises and agrees on decade water withdrawal limits for the Amudarya River in light of water consumers for a particular period and submits them to its territorial administrations for implementation after water withdrawal limits for each state are approved at ICWC meetings.

It should be noted that based on forecasts and given the water-related situation in the region, the following water distribution options are discussed at ICWC meetings:

1. Water is shared according to approved water withdrawal limits without reduction during a normal year and available storage in reservoirs.
2. Provision in Article 4 under Almata Agreement on Central Asian countries of 18.02.92 is used during low-water periods. The following criteria are established for interstate use of fixed water withdrawal limits:

- When water availability is lower than the designated level, water withdrawals for countries should be reduced proportionally within the whole river basin according to an ICWC decision;
- There should be no more than 10% withdrawal above the established limit for certain periods of time;

- BWO "Amudarya" puts in place percentage water sharing between water consumers under conditions of water deficit in the river basin for certain periods of time.

At the same time, BWO is based on the following principles of percentage water sharing between the states in the Amudarya River basin: when establishing a share of water withdrawal under percentage water sharing, water withdrawal limits are approved by ICWC for key water consumers for the whole period. There are no problems in surface water management and distribution under conditions of sufficient water availability in the basin. BWO and ICWC immediately address problems that emerge during irrigation periods. Water resources management is complicated during low-water periods in extreme events, when ICWC makes decisions on water withdrawal limitations which are obligatory for all water consumers. In spite of strengthened joint control by BWO and ICWC, these decisions are not always fulfilled due to imperfect legal and regulatory frameworks of ICWC executive bodies, different national interests in water use and poor equipping of BWO structures with water accounting and monitoring facilities.

One of the main principles for the Amudarya River water resources management and regulation at the BWO level is settlement of flow delivery and receiving within border stations and introduction of in-system and streamflow balance in waterworks management. The border stations are Kelif, Darganata, Tyuyamuyun, Kipchak and Samanbai. At present, BWO makes regular balance calculations within river reaches between the gauging stations Kelif and Samanbai with four balancing sites (Kelif – Darganata, Darganata - Tyuyamuyun, Tyuyamuyun - Kipchak, Kipchak – Samanbai).

Drawbacks of Regional Cooperation on Water Resources Management in the Amudarya River Basin

The following organizations are involved in regional cooperation of water resources management in the Amudarya basin:

1. From IFAS:
   - ICWC
   - BWO "Amudarya"
   - SIC ICWC
   - ICWC Metrological Center
2. UDC "Energy"
3. From states – hydrometeorological services.

The accepted form of water resources management in the Amudarya basin is fully applicable, but it has a number of serious functional problems:

1. It should be noted that there are still no conditions for successful and effective implementation of management and cooperation mechanisms in the region:

- While recognizing ICWC and BWO as the key units in water management and allocation at the regional level, their current status is obviously undervalued and does not facilitate the successful implementation of set objectives.
- ICWC and BWO capacities are not used fully. In particular, ICWC recommendations to ensure agreed upon conditions for water distribution and releases to the Aral Sea are not always observed.
- One of the key drawbacks in ICWC activity is the fact that its decisions do not always reach superior bodies in CAR countries.

2. BWO powers as an executive body on interstate water distribution are limited due to the following reasons:

   - Some transboundary water diversion structures, as well as the most significant hydropower systems with reservoirs are managed by national bodies, and not by BWO;
   - BWO does not control volumes and schedules of ground water withdrawal and return water disposal or water resource quality;
   - Protection zones of interstate rivers are still not defined.
   - Respective reaches of the Amudarya and other rivers are under the jurisdiction of national bodies and BWO does not practically control the situation within such reaches.
   - Interaction between BWO and national hydrometeorological services is not coordinated. This impacts negatively on the accuracy of available water accounting and forecasting.
   - BWO and its subordinate bodies do not have a sufficient technical base for information acquisition, processing and communication.
   - The river channel and protection zone are not monitored and controlled.

Thus, the existing water resources management structure for the Amudarya River basin should be strengthened and improved.

Conclusions

In order to achieve a sustainable environmental situation, rational use of water resources and their protection from depletion and pollution is of exceptional importance to the economic, social and cultural development of nations living in the Aral Sea basin. Existing problems related to the hydrology of arid zones in the region are typical of the whole world. They can be of five dimensions:

1. problems related to scientific rationale of optimal water resources use under water deficit;
2. hydrological education and training;
3. creation of information systems;

4. re-equipping;
5. regional cooperation.

Taking into account that water is a natural resource, a product of nature and the public, and a source of impact on ecological and social environments, the scientific field of the problem should be focused to allow the development of a successful water resources strategy, i.e. a set of principal regulations, norms and guidelines. This would be implemented as interstate agreements, contracts and other documents, as well as organizational and economical arrangements and could be tools for sustainable water development in the region, taking into account the particular importance of water resources under arid zone conditions, water deficits and risks of conflict as a result of the significant impact of water on life in the region. The legal mechanism developed using international experience, and joint water management in the Aral Sea basin will serve as a basis to settle many water-related contradictions in Central Asia at the regional and national levels. Sustainable economic development in the region depends on effective interstate cooperation and requires improvement of the legal framework in the area of water relations. It should be pointed out that when studying issues related to regional cooperation, one should consider that practically all international conventions and agreements on water resources use recognize the following:

- Water resources are common property and the basis for future development, and water resource volume is very limited.
- Water resources do not depend on state boundaries.
- The key goal of water resources management is common to nations and countries.
- Basin-wide interests take priority over private ones, including water using countries.
- Obligatory observation of the equitable and rational water use principle and "do not harm" rule

It is a principal provision of the International Waterways Law that any government should use the international waterway in a rational and equitable way with respect to other states using the waterway.

The developed institutional framework for IFAS and its important real capacities can assure sustainable shared interstate water resources management and effective and rational use in the region. To make the IFAS management structure effective, countries should make the necessary efforts and take joint action to attain priority objectives directed at strengthening interstate cooperation in water resources management in the region. Such priority objectives are:

1. Adopting a number of interstate Agreements on institutional framework for shared water resources management and use under current conditions, on information exchange in the region, and on water supply limits to the Aral Sea and Priaralie.
2. Strengthening ICWC powers.
3. Enhancing the BWO "Amudarya" role.

4. Adopting a special Agreement on key principles for transboundary water sharing in the Amudarya River basin.

Security Issues in the Amudarya Basin at the Regional Level

Based on 20 years of experience at BWO "Amudarya", the following important security-related issues should be pointed out in the Amudarya basin at regional level:

1. Technical conditions of river water diversion structures, interstate canals with their structures, supply canals managed by BWO as well as the technical state of reservoirs, river pumping and gauging stations, water diversion structures managed by riparian states and controlled by BWO. BWO "Amudarya" emphasizes this issue. Based on availability of finances from country founders, BWO regularly maintains structures and canals with waterworks transmitted for temporary use. However, it should be emphasized that the allocated funds are sufficient only for maintenance and there are no finances to replace outdated and depreciated structures.
2. Reliability of forecasts provided by hydrometeorological services. There were practically no forecasts for the Amudarya basin during recent years, since hydrometeorological services lack required data. Such a situation creates difficulties for planning and making effective decisions.
3. Floods. They contribute to security issues in the Amudarya basin. Safe passage of flood discharge is a very difficult water-related task. BWO "Amudarya" jointly with riparian states have experience in this respect.
4. Ice phenomena in the river channel and interstate canals.
5. Water shortages in the basin.

Conclusions

During the last year, the world community and the states of Central Asia have paid more attention to improving security and water productivity. As economies grow and stability is improved in riparian states, issues related to cooperation in economic and water aspects and food and water security in Central Asia will gradually be addressed.

Threats to Water Security in the Republic of Kazakhstan: The Transboundary Context and Possible Ways to Eliminate Them

Anatoliy Dmitriyevich Ryabtsev

The Strategy for Industrial Innovation Development in the Republic of Kazakhstan by 2015 has an objective to improve the efficiency of material consumption in the national economy. This objective relates, along with other kinds of resources, to water, which is of strategic importance as it directly influences the social component of the national economy's competitiveness – population health. Kazakhstan is one of the states that has limited water resources and experiences severe water scarcity for satisfying the needs of the economic sectors as well as for providing the population with drinking water.

It is generally recognized that energy resources such as oil and gas are considered strategic resources. Having them available enables a certain influence in world economy and politics. However, this issue should be considered only in the context of any given region. For example, according to the "Economist" magazine, water resources and their availability will play a key role in the Central Asian region in the near future. It is explained that oil and gas are exhaustible energy sources, while water resources are renewable.

Most of Kazakhstan's territory comprises desert and semi-desert where water supply is a very acute problem in terms of not only economic activity, but also consumption by the population. The problems of sustainable water supply and water security in Kazakhstan are severe because the most important surface water sources are located in neighboring Russia, China, and Central Asian countries. In an average wet year, a total of about 44 km^3 of water comes into Kazakhstan through such transboundary rivers as the Ural, Black Irtysh, Ili, Chu, Talas, Syrdarya and others. Total available water resources of the country amount to 100.5 km^3.

Expansion of economic activity in these countries leads to a reduction in inflow of water resources to the territory of Kazakhstan. This is an interstate problem that requires the development of an appropriate regulatory legal mechanism. It would be valuable to examine the experience of, for example, European countries on the use

A.D. Ryabtsev (✉)
Committee for Water Resources, Ministry of Agriculture,
9 Avenue Pobeda Astana, 010000, Republic of Kazakhstan
e-mail: cwr_mac@mail.ru

C. Madramootoo and V.A. Dukhovny (eds.), *Water and Food Security in Central Asia*, NATO Science for Peace and Security Series C: Environmental Security,
DOI 10.1007/978-90-481-9974-7_6,

of water resources in the Danube and other rivers. At present, interstate relationships with adjacent countries with regard to shared use and protection of transboundary water resources are established on the basis of the agreements currently in force.

Within Central Asia

The relationships among Central Asian countries with regard to shared use and protection of transboundary river water resources are regulated by several basic agreements, which have been rendered competent by the Heads of Central Asian states in a number of documents, including in the Nukus Declaration of Central Asian States and International Organizations on Sustainable Development in the Aral Sea Basin adopted on 20 September 1995. All of them should be put into effect in a steadfast manner.

1. Agreement between the Republic of Kazakhstan, Kyrgyz Republic, Republic of Uzbekistan, Republic of Tajikistan, and Turkmenistan on Cooperation in the Field of Joint Management and Protection of Interstate Water Resources signed on 18 February 1992 (Alma-Ata).
2. Agreement on Joint Action for Solving the Aral Sea and Priaralie Problem, Environmental Sanitation and Socio-Economic Development in the Aral Region signed 26 March 1993 (Kzyl-Orda).
3. Agreement between the Governments of the Republic of Kazakhstan, Kyrgyz Republic and Republic of Uzbekistan on the Use of Water-Energy Resources in the Syrdarya River Basin signed on 17 March 1998 (Bishkek).
4. Agreement between the Governments of the Republic of Kazakhstan and Kyrgyz Republic on the Use of Interstate Waterworks Facilities on the Chu and Talas Rivers signed on 21 January 2000 (Astana) (ratified by the Law #301-II of the Republic of Kazakhstan of 7 March 2002).

Along with these agreements and the 1998 Framework Agreement, bilateral and multilateral inter-governmental Protocols are signed every year which stipulate terms of compensatory delivery of water, electric energy and fuel between the countries. For the years 2005, 2006, and 2007, Uzbekistan has not signed, at its sole discretion, a quadrilateral Protocol on the Use of Water-Energy Resources in the Naryn-Syrdarya reservoir cascade (between the governments of Kazakhstan, Kyrgyz Republic, Tajikistan and Uzbekistan). Thereupon, the same bilateral Protocols are signed every year between Kazakhstan and Kyrgyz Republic, and between Uzbekistan and Tajikistan. The bilateral Protocol does not guarantee Kazakhstan a water supply in the Syrdarya River up to the Shardara reservoir in such quantity, for equivalent of which the Kyrgyz electric energy is purchased. This is because the share of the Syrdarya River water resources, which is allotted for Kazakhstan, is partly used in Uzbekistan and Tajikistan.

In the near future, the following steps should be taken:

1. Developing and signing a long-term agreement between the Governments of the Republic of Kazakhstan, Kyrgyz Republic, Republic of Tajikistan and Republic of Uzbekistan on the Use of Water and Energy Resources in the Syrdarya River basin. The draft Agreement has mainly been developed within the framework of the Asian Development Bank RETA "Improved mechanism for management and regulation of water resources in the Amudarya and Syrdarya River basins". After signing, this Agreement will replace the 1998 Agreement and should ensure adjustment of operation modes for the Naryn-Syrdarya reservoir cascade through long-term flow planning and regulation, coordination of yearly water release regimes, electric power generation and transmission, and energy resources compensation on a contractual basis. Furthermore, it will assign to each Party a share of limits on water withdrawals from the Syrdarya River channel for the hydrological year before the confirmation of a new strategy for water distribution in this river basin. It also establishes a mechanism for providing water-energy regimes for the Naryn Syrdarya reservoir cascade, stipulates issues regarding joint consideration of the construction of new hydroelectric schemes and reservoirs in the region, development of large irrigated land areas, economic mechanisms with regard to international water relations, water conservation, reduction of polluted water discharge into water bodies, exchange of information and forecasts and so on.
2. Developing and signing, within the framework of the Eurasian Economic Community (EurAsEC), a Concept for Effective Use of Water-Energy Resources in Central Asian region and then Agreements on the Creation of International Water-Energy Consortium in Central Asia. The Concept is a set of agreed views and approaches to principles for interaction of EurAsEC member states in joint development of hydroelectric potential and effective use of water-energy resources in the Syrdarya and Amudarya River basins. It determines favorable economic and legal conditions for economic entities in water, fuel-energy and other sectors of EurAsEC member states in this field. The provisions of the Concept serve as a basis for development of an interstate Agreement and other legal acts with regard to water-energy resource use in the Central Asian region.

In Kazakhstan, along with the work on interstate cooperation, development of the Kazakh part of the Syrdarya River is being done within the framework of the "Regulation of the Syrdarya River Channel and Conservation of the Northern Aral Sea" Projects 1 and 2. Thanks to the implementation of the "Regulation of the Syrdarya River Channel and Conservation of the Northern Aral Sea" Project 1, all the major hydraulic structures on the river were reconstructed, new structures were constructed, river-channel straightening work was done at several sites, and check dams along the river channel were restored. As a result of these measures, the Syrdarya River winter flow capacity increased from 350–400 m^3/s to 700 m^3/s, which partly solved the inundation problem of riverside territories during man-made winter floods. The northern part of the Aral Sea was separated by a dam and filled with water from the Syrdarya River.

Within Phase 2, deltaic lakes will be restored, and further impoundment of the former seabed in Saryshyganak bay will be carried out. Phase 2 also includes the construction of a Koksarai counter-regulator, which will enable the complete elimination of emergency situations on the river during the winter period, when there is no opportunity to discharge flood water into the Arnasai depression. To date, owing to the construction of the Arnasai reservoir, releases from Shardara have declined by almost 4 times from 2,600 to 600 m^3. This has begun to cause a great danger: loss of the sustainability of Shardara's main dam. During a flood, the water body can quickly be overfilled. There was a very dangerous incident in the history of the Shardara reservoir. In 1969 a large-scale flood lasted until the middle of summer. Flows reached 2,000 m^3/s and only the complete opening of the sluice gates prevented a catastrophe. During the development of Phase 2, the following objectives were set. Firstly, it became necessary to ensure the sustainability of the Shardara dam and, essentially, conserve the reservoir. Secondly, and what is probably more important, is the elimination of the threat of emergency situations for settlements in South Kazakhstan and Kzyl-Orda provinces. Thirdly, it is necessary to improve irrigation water supply to lands in Kzyl-Orda province, provide guaranteed water supply to fill deltaic lakes in the lower reaches of the river, and keep the necessary size of the Small Aral. Moreover, there is another very important factor. One cannot disregard that Kyrgyzstan is gradually winning the Central Asian electric power market and plans to enter the international markets of Pakistan and Afghanistan. Kyrgyzstan is installing new generation capacity on its hydroelectric power stations. This will lead to the discharge of additional water, especially in the winter period, from the Naryn-Syrdarya hydroelectric stations cascade. In this case, the analogue of Arnasai in the Kazak territory can generally relieve the tense water situation in the Syrdarya River downstream reaches.

People's Republic of China

Water relations between the Republic of Kazakhstan and the People's Republic of China are also being established on the basis of agreements signed owing to the joint work. The basis for interstate cooperation in the field of water relations between the two countries is the Agreement between the governments of the Republic of Kazakhstan and People's Republic of China (PRC) on the Cooperation in the Use and Protection of Transboundary Rivers signed on 12 September 2001 (Astana). It should be noted that earlier PRC concluded such agreements only with Mongolia and Korea. It is only recently that a similar agreement between PRC and the Russian Federation has been concluded. Currently, trilateral cooperation among China, Kazakhstan and Russia in water resources use and protection in the Irtysh river basin is not accepted by the Chinese party.

In order to execute Article 8 of the above-mentioned Agreement, a *Kazakh-Chinese Joint Commission on the Use and Protection of Transboundary Rivers* has been founded. The main result of the Joint Commission's operation is the signature of several agreements:

- Agreement between the Ministry of Agriculture of the Republic of Kazakhstan and the Ministry of Water Resources of the People's Republic of China on emergency notification of the Parties of Natural Disasters on Transboundary Rivers signed 4 July 2005 in Astana;
- Agreement between the Ministry of Agriculture of the Republic of Kazakhstan and the Ministry of Water Resources of the People's Republic of China on Development of Scientific-Research Cooperation on Transboundary Rivers signed 20 December 2006 in Beijing
- Agreement between the Ministry of Environment of the Republic of Kazakhstan and the Ministry of Water Resources of the People's Republic of China on Exchange of Hydrological and Hydrochemical Information (Data) of Border Gauging Stations on Major Transboundary Rivers signed 20 December 2006 in Beijing

Until recently, the fundamentally important issues for Kazakhstan that are related to control over water quality in transboundary rivers and prevention of their pollution, as well as the consideration of the principle for water distribution along transboundary rivers were not properly understood by the Chinese party. The last meeting of the Joint Commission held in Beijing in December last year succeeded in moving ahead with regard to these issues. A draft was submitted to the Chinese party on the Concept for Water Distribution along the Irtysh and Ili Rivers, and consideration of the draft Agreement on control over transboundary water quality and pollution prevention, prepared by the Kazakh party, was achieved.

In the near future, the following steps should be taken:

1. Accelerate the signature of agreements:
   - On cooperation in environmental protection
   - On control over water quality in transboundary rivers and prevention of their pollution
   - On the principles of water distribution on the Irtysh river
   - On the principles of water distribution on the Ili river

   The position of Kazakh specialists on water distribution should, first of all, be based on the ecosystem approach. Socio-economic requirements for water resources should be taken into account in the complex, not infringing requirements of the environment.
2. Continue the efforts for preparing and concluding a trilateral agreement (Russia, Kazakhstan and China) on cooperation in the field of water resource use and protection in the Irtysh River basin, and in particular use the format of measures carried out within the framework of the Shanghai Cooperation Organization. It is necessary to use potential contacts in all forms: diplomatic channels, meetings of national government officials, resources of the intergovernmental agreements in force, other formats.

3. Accelerate the construction of joint hydroelectric schemes on transboundary rivers (Khorgos and other rivers) that will prevent the interception of a large quantity of water by China and guarantee water supply to its users in required quantities.
4. Urgently provide laboratories and gauging stations in Kazakhstan with modern equipment and measurement instrumentation to improve the precision of measured parameters and study the chemical composition of water in transboundary rivers. The postponement of solving these issues will lead to further pollution of inflowing water resources, and defer *sine die* the conclusion of an agreement on water quality.

Russian Federation

At present, the Agreement between the governments of the Republic of Kazakhstan and the Russian Federation on Shared Use and Protection of Transboundary Water Bodies signed on 27 August 1992 (Orenburg city) is the basis for interstate cooperation with regard to water relations between the Republic of Kazakhstan and Russian Federation. In order to execute Article 11 of this Agreement, a Joint Kazakh-Russian Commission on Shared Use and Protection of Transboundary Water Bodies was founded on parity basis. Within the framework of the Commission, working groups were established on transboundary Ishim, Irtysh, Tobol, Ural, Big and Small Uzeni River basins.

In recent years, one of the topical issues of interstate cooperation among our countries is the issue related to the competence of declaring irrigation water supplied from the Pallasov irrigation system in Volgograd province (Russia) to the Janibek irrigation system (Kazakhstan), which was raised by the Russian party. The issue has not been solved yet. The other topical issue to be solved is a trilateral Kazakh-Russian-Chinese cooperation in shared use and protection of transboundary water resources in the Irtysh River basin. The parties deem it necessary to continue joint actions to mitigate and change the position of China on this issue, using potential contacts in all forms, including diplomatic channels, meetings of national government officials, and resources of the acting intergovernmental agreements of other formats.

In the near future, the following steps should be taken:

1. Conclude a new long-term Agreement, considering the experience of cooperation accumulated since 1992, based on the rules of international water law.
2. Sign an inter-governmental agreement on exemption from water tax, customs duties for custom registration and declaring of water supply from the territory of one country for environmental and irrigation needs of another country.
3. Study the issues of preparing and signing a trilateral agreement on cooperation in the field of shared use and protection of transboundary water resources in the Irtysh river basin together with the Chinese and Russian parties.

None of these steps to retain water security in the Republic of Kazakhstan will be successful if regular and purposeful work is not done on an extensive introduction of resource and water-saving technologies in enterprises, water reuse in industry, new and more rational irrigation technologies, raising broad public awareness and the need for a respectful attitude to water as a valuable and universal natural resource. All this work should be done within the framework of the Program for integrated water resources management.

The Role of Land Reclamation in the Socio-Economic Progress in Uzbekistan and Government Support Policies of Water Management

Shavkat R. Khamraev

Uzbekistan has a long history of irrigated agriculture. At present, 65% of the Uzbek population lives in rural areas and 44% of the working population is employed in agriculture. The contribution of agriculture to the country's economy is considerable: the share of the agricultural sector in the GDP ranges from 30% to 40% in different years.

In 2007, the growth rate of agriculture remained at previous high levels and amounted to 6.1%. This is considered a fairly strong rate even for leading economies of the world which have efficient agrarian sectors. This growth has been achieved due to strong government support of agriculture in recent years and through the use of a wide range of financial instruments including: tax preferences, preferential financing of the whole production cycle (including sowing and harvest campaigns), broad implementation of leading scientific research results of Uzbek scientists, establishment and development of service infrastructure, development and implementation of government programs on improvement of lands, and equipping with advanced agricultural facilities and machines.

Uzbekistan ranks fifth among 90 cotton-growing countries in cotton production and second in cotton fiber exporting. The large export potential of processed cotton fiber products has resulted in the establishment of large textile companies and a clothing industry, including joint ventures with foreign investment.

Uzbekistan occupies one of the leading positions in CIS in the production of fruit, grapes, cocoons, karakul and wool. The created material and technical base generates fresh and canned produce in volumes that meet the demands of the population as well as export demand.

Major structural and institutional transformations occurred during agrarian reforms in rural areas. State-owned agricultural enterprises were reorganized into different private enterprises. Besides structural changes in land resource reallocation and crop production, cotton planting was reduced and stabilized at 1.5 million ha.

S.R. Khamraev (✉)
Ministry of Agriculture and Water Resources of Uzbekistan,
4 Navoiy Str., Tashkent, 100004, Republic of Uzbekistan
e-mail: info@agro.uz

C. Madramootoo et al. (eds.), *Water and Food Security in Central Asia*,
NATO Science for Peace and Security Series C: Environmental Security,
DOI 10.1007/978-90-481-9974-7_7,

Since independence, private farm areas have increased by 150,000 ha, and now account for more than 650,000 ha. The problem of grain self-sufficiency was addressed and a stable production of over 5 million tons of produce, grapes and potatoes has been achieved.

As was mentioned in the 2007 report of the president of the Republic of Uzbekistan I.A. Karimov, in previous years, particular consideration was given to the development of agriculture and the country has recently experienced a revolution in agriculture due to stable reforms in this sector. The production of high volumes of raw cotton and grain in 2007 (over 3,650,000 t of raw cotton and 6,250,000 t of grain), provides evidence of the above mentioned change in agriculture. Today there are more than 215,000 farms, which employ 1.6 million workers. Moreover, about 85% of all planted areas in agriculture are assigned to them. The share of farms in cotton and grain production represents almost 99% and 82%, respectively. All of this became possible due to the development of new market infrastructure, establishment and promotion of agricultural production, and changes in principles and management systems of the sector.

Development of irrigation in the previous period defined by wasteful use of irrigation water brought about certain changes that contributed to raising the level of groundwater over large territories. Under arid conditions (as found in Uzbekistan), this caused an increase in salinization and a decrease in land productivity. To avoid the negative impact on land productivity, construction and reconstruction works of horizontal and vertical drainage systems were carried out.

Irrigated Land Salinity

The salinity of lands in the Republic was caused by reclamation of areas naturally exposed to salinization as well as secondary salinization caused by an increase in the level of alkaline soil water. Decreasing productivity of irrigated lands in recent years, invoked partly by deteriorating irrigation and land reclamation infrastructure, has aggravated and worsened salinity and caused expansion of saline lands.

Saline irrigated lands in the Republic constitute 2,170,900 ha or 50.7% of total irrigated lands and within this area moderate and highly saline lands amount to 826,400 ha (19.3%). The areas prone to salinization are mainly located in the lower reaches of the Amudarya River and in the Sirdarya River basin (Sirdarya and Djizak oblasts and Central Fergana). From 1995 to 2004, areas of moderately and highly saline lands in the Republic increased by 14.4% in the Djizak, Kashkadarya, Namangan, Sirdarya, Fergana and Khorezm oblasts. The worst water logging is observed in the Khorezm oblast, where land with a high groundwater level exceeds 80% of total land given their low salinization (less than 3 g/L). When used for agricultural purposes, a seasonal salinization of irrigation lands occurs. The situation is critical in Khorezm, whereas it is slightly better in Karakalpakstan: lands with shallow groundwater account for a little more than 20%. The main reason for the increase in groundwater level is large water losses from canals and fields caused by poor land planning and irrational use (losses) of water in the field and inadequate water diversion from irrigated lands.

Seasonal salinization is eliminated by autumn–winter washing, which has a low efficiency due to poor operation of drainage and water diversion systems of collectors.

Availability of Drainage in Irrigated Lands

Drainage must be built on 3.31 million ha out of a total irrigated area of 4.2 million ha; 2.93 million ha of land is provided with collector-drainage network. Demand for drainage by oblasts of the Republic ranges from 40.6% in Samarkand oblast to 100.0% in Bukhara, Sirdarya and Khorezm oblasts. Horizontal drainage is built on 91.3% of irrigated lands and vertical drainage on 8.7%. The actual specific length of drainage on irrigated lands in all oblasts, except Fergana, is much less than recommended.

To ensure the ameliorative condition of irrigated lands, 138,800 km of drainage network is operated in the Republic, and of this 29.5 % is operated by the state and 70.5% by land users depending on the availability of the network. The network is 83% in the Republic of Karakalpakstan, and 88% in the Kashkadarya oblast. Farmers operate on-farm collector-drainage networks under new conditions and delegate these functions to established water user associations (WUAs).

The state of irrigated lands in the Republic is predetermined by the technical condition of all infrastructures, composed of land reclamation systems from the main diversion network to on-farm drainage, irrigation regime, adherence to techniques and required volume of washing irrigation.

The Decree of the President of the Republic of Uzbekistan « On measures of fundamental improvement of land reclamation system » of October 10, 2007 states that the present condition of irrigated lands is slowing down further growth in crop yields and revenues of agricultural product manufacturers. The lack of a complex and systematic approach to the development of land reclamation policies, concrete sources of financing and weak performance of water organizations and WUAs has lead to a reduction in the scope of land reclamation works in recent years and to an increase in salinization and groundwater levels. Besides economic damage and loss to farmer profits, salinization and water logging create ecologically unfavorable conditions (flooding of residential areas, low quality of drinking water), negatively affecting living conditions and people's health. The abovementioned decree defines a cardinal improvement in the condition of irrigated lands as the most important priority of agricultural development for 2008–2012.

The land reclamation program for 2008–2012, developed in accordance with the above Decree ensures the fulfillment of primary measures for the development and improvement of irrigation and drainage systems and for increasing the efficiency of drainage systems operation directed towards improving conditions, from budget sources. The program consists of the followings tasks:

1. Construction and reconstruction of 3,529 km of main and on-farm collectors, 1,031 vertical drainage wells, 107 pumping stations for land improvement, restoration of 7,600 km closed horizontal drainage.
2. Repair and renewal works along the collector systems with an overall length of 126,700 km, including all infrastructure located in the operating zone: vertical drainage wells, reclamation pumping stations, hydraulic facilities and culverts, etc.
3. Replacement of current stock and purchase of 624 units of new land reclamation technologies for increasing the material-technical base of construction and operating entities for carrying out works on the improvement of land reclamation of irrigated lands.

Preliminary costs, determined by the program on land reclamation and improvements for 2008–2012, amount to 604,163.2 million sum.

The Ministry of Agriculture and Water Resources acts as the Central Water Administration and is responsible for water resources management, planning and allocation of water across sectors in the economy and oblasts of the Republic. Due to the limited water resources in the region, water sector specialists are working on the implementation of water saving technologies, drip irrigation systems and achieving scientific and technological progress. The guaranteed supply of good quality water to irrigated agriculture and all sectors of the economy is one of the main goals. Water resources make a huge contribution towards achieving planned yields of cotton, grain and other agricultural products. More than 95% of crops in the Republic are obtained from irrigated arable land.

According to the Resolution of the Cabinet of Ministers of July 21, 2003 « About water management improvement » all secondary water users, located in the territories of dismantled shirkats (agricultural cooperatives) have been incorporated into water user associations. The following tasks are assigned to these groups: equal use of on-farm irrigation networks, equitable water allocation between farmers and water users, implementation of advanced irrigation technologies, and protection of association member rights.

The volume of WUA services to secondary water users in 2007 that supply water to an area of almost 4 million ha is over 16 billion sums. Of total water consumed in Uzbekistan, 92% is used for irrigation and therefore much attention is given to conservation. As a result of measures taken, 4.6 billion m^3 of water was saved in 2007 compared with 2006.

Since the water sector incurs high costs, in the first years of independence, the government could not support this sector due to economic reasons. From 1991 to 2001, the share of government investment in agriculture declined from 27% to 8%. About 20% of the electricity is consumed in the country and 70% of the MAWR budget is allocated to electricity used by pumping stations and drainage. Allocation of funds for maintenance and operation of the irrigation and drainage infrastructure have sharply decreased, repairs as well as cleaning of collectors and drains have declined, and reconstruction of canals and hydraulic facilities has stopped. At present, wearing out of basic funds account for 30–50%. The reliability of irrigation services per hectare of land in a normal year is about 30%, indicating significant problems in the operation of irrigation systems. Overall requirements for water infrastructure rehabilitation may be summarized as follows:

- 32.1% of all on-farm and main canals (22,300 km) need reconstruction and repair (23.5%)
- More than 42.1% of on-farm irrigation network (149,500 km) need reconstruction, while 17.4% need repair
- Out of 42 water withdrawal intakes with a capacity ranging from 10 to 300 m^3/s, 18 require replacement and restoration of hydromechanical equipment; five water works require reconstruction
- Eighty percent of big, 50% of medium and 30% of small pumping stations require repair and reconstruction
- Sharp increases in the prices of utilities and equipment have changed priorities for the benefit of self-flowing irrigation
- About 19,000 km of farm open drains require cleaning, 11,500 km of open and closed drains need reconstruction and repair, and no more than 50% of closed horizontal drainage systems are in working order

Water Management Problems

Stable development of irrigated agriculture may be threatened by a possible water resources deficit due to:

- Water supply fluctuations caused by a lack of water policy coordination among Central Asian countries and inadequate run-off regulation within the republics
- Inadequate water resources management within main river basins
- Considerable amount of unproductive water losses in the irrigation network during transportation
- Inefficient use of water in the fields during irrigation

Ways of Saving Water

1. The technological level and efficiency factor of irrigation systems should be improved through their reconstruction. The technological level of systems, which determines the extent of losses and the efficiency factor, is specified by material and technical possibilities of the Republic. Under present conditions, at the optimal level of technical condition, the efficiency factor of the systems should reach 0.80–0.83. Such an indicator can be produced by cutting the length of the on-farm irrigation network and implementing economically advanced canal construction (concrete, flume and pipeline).

2. Increase in irrigation efficiency. Furrow irrigation will remain important in Uzbekistan in the near future, due to several reasons: adherence of farmers to traditions, high-energy intensity and duration of advanced water saving technologies implementation. It is possible to save a significant amount of water if adequate measures are taken on steeply sloped lands and mildly sloped lands, which account for 1,511,400 ha or 38% of all irrigated area of the Republic. Improving infrastructure in district irrigation networks on these lands should be the primary task. In the allocation of water within the furrows, structural solutions consisting of flumes, buried pipe systems (pressure and non-pressure), automated flumes and buried static pipes play an important role in the improvement of irrigation techniques. Improvement of techniques on mildly sloped lands requires accurate, thorough and practical planning of lands, construction of fine network of flumes or buried pipelines, and application of flexible pipes, as well as automated flumes. Drip and subsoil irrigation are methods for orchards and vineyard irrigation, respectively, on very steeply sloped lands. However, difficulties in usage and lack of financial mechanisms to increase farmer interest in the implementation of these irrigation techniques prevent wide dissemination of these methods.

Conclusion

In recent years the government has paid a great deal of attention to the improvement of water sector performance. Firstly, it is concerned with the improvement of land reclamation. In subsequent stages of water management development, attention must be given to measures aimed at increasing rational water use: reducing water losses during transportation and in the fields.

With insufficient financial resources it is important to solve the dilemma of where to invest first: in enhancement (rehabilitation) of irrigation infrastructure or in improvement of water use culture in the fields. Obviously, these closely related issues should be considered and solved simultaneously within concrete systems and on the basis of technical and economic analyses.

The program of evolutionary development and implementation of water saving technologies is being developed in the Republic. To identify stages of the program, it is necessary to understand where the largest amount of water is being lost and where ecology is being damaged most. Territories that require reduced seepage losses in the fields as a priority are:

- Lands where water is costly: pumping irrigation lands: 1.53 million ha.
- Highly water permeable lands: 1 million ha (including lands with low silt depth underlain by gravel).
- Automorphic lands, which have deep groundwater and relatively shallow groundwater –0.5 million ha. These are so called « adirs», – on which, application of furrow irrigation, leads to degradation of lands in the lowlands (water logging and salinization).

Since water-saving technologies have high purchasing and operating costs, the following requirements should be met to implement them:

- Profitability of agricultural production
- Promotion of users in order to cover operating costs (for example reduced tariffs for electricity)

In addition, advocacy, education and the possibility of purchasing equipment should be ensured. As a first stage, the use of drip irrigation for orchards and vineyards seems quite reasonable. In order to implement water saving irrigation technologies, investments or subsidized government loans (other sponsor loans), as well as legal, financial mechanisms and technical support (production and possibility of implementing the irrigation techniques) are needed.

The experience of cooperation with international financial institutions in the implementation of joint projects on improving the reclamation of irrigated lands in the Akaltin district of the Sirdarya oblast, southern regions of the Republic of Karakalpakstan and rehabilitation of irrigated lands in nine districts of Bukhara, Kashkadarya and Navoi oblasts have fully paid the cost. The World Bank and SCO subsidized loans amounting to around 70 million USD were obtained to finance the realization of a project to take drainage water away from South Karakalpakstan and improve the condition of the Central Fergana lands.

In the report, devoted to the results of 2007 and tasks for 2008, the President of the Republic of Uzbekistan stated the importance of promoting activities to further attract foreign subsidized loans by the Ministry of Agriculture and Water Resources, Ministry of Economy, Ministry of Finance, Ministry for Foreign Economic Relations, Investments and Trade for implementing projects in this sphere.

Experience of Coordinated Water Resources Use of the Syrdarya River Basin States

Makhmud Khamidov

The Syrdarya River basin covers an area of 4,85,000 km^2 in the territory of four sovereign states: the Kyrgyz Republic, Uzbekistan, Tajikistan and Kazakhstan. At present, more than 25 million people live on the banks of the Syrdarya River and use the majority of the available water resources for their subsistence and development.

The Syrdarya River water resources amount to more than 40 km^3, more than 60% of which come from the highlands. The flow of the river and its main tributaries are regulated by the Naryn-Syrdarya reservoir cascade and distributed using hydro schemes, intakes and canals. The total effective capacity of the cascade's reservoirs amounts to 24 km^3, and the nine major hydroelectric power stations have a total installed capacity of 3,720 MW. These structures were constructed mainly in the middle of the last century and over the past 30 years have enabled a 250% increase in agricultural production and provided employment for a major part of the growing rural population constituting 60% of the population living in the Syrdarya River basin.

As hydro schemes and reservoirs were put into operation on the Syrdarya River and its tributaries, a complex engineering system, functioning under conditions of rapidly growing water consumption and limited water resources, was developed. Difficulties with water resource management in the basin were inevitable. As early as the mid 1970s, under practically complete flow regulation, water deficiencies began to emerge in low-water years. To optimize management in the mid 1980s, an Automated Control Complex in Tashkent was established as well as territorial control offices in Andizhan, Leninabad, Chirchik and Gulistan. These were reorganized into the "Syrdarya" Water Administration in 1987. Through its control of water withdrawals from rivers of the basin into canals in the republics, the Water Administration was charged with helping to reduce water deficiencies arising at that time in the southern provinces of Kazakhstan and removing obstacles to passing water to downstream areas of the Syrdarya River and the Aral Sea.

M. Khamidov (✉)
Basin Water Organization Syrdarya, B.11, Karasu-4,
Tashkent, 700187, Republic of Uzbekistan
e-mail: bvosyrdarya@mail.ru

C. Madramootoo et al. (eds.), *Water and Food Security in Central Asia*,
NATO Science for Peace and Security Series C: Environmental Security,
DOI 10.1007/978-90-481-9974-7_8,

The emergence of such problems resulted in a need to organize the control offices at the height of the growing season. A number of specialists were involved in this organization, which helped to supply water to users and carry out measures for water passage. However, the effectiveness of actions taken by the Water Administration was insufficient since its functions, which were limited to water supply control only, forbade them from directly managing the operation of structures which were under the jurisdiction of the union republics. Therefore, BWO "Syrdarya" was founded in early 1988 and was given major hydro schemes and intakes for temporary use. In the first years of its operation, BWO "Syrdarya" ensured observance of limited water supply and appropriate accounting of river water resources. The effectiveness of management and prompt regulation of water resources improved, and water losses were reduced to 1 km^3. The water user republics began to regularly receive water quantities foreseen by limits, and the required inflow to the Chardara reservoir was provided.

After the emergence of five sovereign states in Central Asia, the Syrdarya River turned into an international watercourse according to current terminology and it became necessary to coordinate water resource management of the riparian countries. To this end, an Interstate Commission for Water Coordination (ICWC) was established in early 1992, which lead the work on water resource management in the Aral Sea basin. In the 1992 Alma-Ata Agreement, water sector leaders, on behalf of the five Central Asian national governments, proclaimed their interest in strengthening and developing cooperation in water use, further improvement of water resource sharing, and their adherence to the joint transboundary water management principles. Special attention to environmental protection and the Aral problem was highlighted. BWO "Syrdarya" became an ICWC executive body to implement its decisions and directly manage water resources in the Syrdarya River basin. At present, BWO organizes water supply to ICWC member states, conducts the operation of hydro schemes and intakes, and carries out measures for improving the environmental situation and controlling the quality of used water resources. BWO controls the regime in the Naryn, Karadarya, Chirchik and Syrdarya Rivers from the Uchkurgan and Andizhan hydropower stations to the Chardara reservoir. BWO also exercises online dispatch control of water and energy (together with UDC "Energy") resources in the Syrdarya River basin, and operational control over the observance of water withdrawal limits and electric power overflows. Every month BWO presents ICWC members with full information on current water resource use.

Having transferred major interstate hydro structures to BWO for temporary operation, the Syrdarya River basin states assumed obligations to finance the BWO's activities based on share holding in proportion to water quantities used. It was determined that all hydrological infrastructure of BWO, wherever it is located, directly or indirectly ensures equitable water resource distribution between the countries. Therefore, each participant makes a certain general contribution to financing operation and maintenance.

At the disposal of BWO are head intakes on the Syrdarya River, its major tributaries and main canals as well as 260 km of interstate canals: the Dustlik and

Big Fergana Canal (BFC). BWO operates 203 hydraulic structures, 21 of which are located on the main channels of the Naryn, Syrdarya, Karadarya and Chirchik Rivers. Their flow capacities vary from 20 to 2,500 m^3/s. In addition, BWO has 165 km of collector-drainage networks, 250 units of motor transport, machines and mechanisms, 35,000 m^2 of office and production premises, and 3,200 ha of water protection zones.

BWO maintains accounting of water withdrawals from rivers and lower canals using 445 structures. They include 21 head intakes to main canals, 36 fixed and 172 temporary pumping plants, and numerous outlets from main canals. Water accounting on intakes from main canals is carried out together with the water authorities of the Central Asian states. Every year BWO controls water resources amounting on average to 34 of 37 km^3 or over 90% of the annual river flow in the Syrdarya basin. ICWC member states have entrusted the implementation of their decisions BWO and indicates the significant responsibility of BWO.

To fulfill the objectives given by ICWC, BWO always tries to support management capacity at a level that ensures solving problems in a timely and high quality way. Since 2001, BWO has introduced automated control and management systems with the support of ICWC. At present, BWO has at its disposal ten such systems on interstate headworks and unique hydro schemes. In addition, a data transmission system is being introduced on structures in the Fergana Valley that will integrate the Uchkurgan and Kuyganyar hydro schemes, the head intake at the Big Fergana Canal on the Naryn River, the Khakulabad distributor on the Feeder Canal, structures on the Big Andizhan Canal, and head and escape structures on the Akhunbabayev Canal into one system of automated technological information transmission. Such automation enables a high accuracy of regulation and maintains assigned discharges, reduces the inaccuracy of water accounting, and improves the quality and lightens the work load of operating personnel.

Through a twofold increase in the accuracy of regulation and water accounting, a reduction in nonproductive water losses is achieved that is especially important during water shortages. Electronic means for receiving, storing and transmitting information included in the automated system provide open access to data of interest. In this way information transparency is ensured, doubts of users about water supply accuracy are eliminated and trust in BWO activities is improved. This helps solve the issues of water resource management in a conflict-free and trusting climate.

In recent years BWO has been working on the creation of information systems, considering that large bodies of different information are used in the management. The information database of BWO "Syrdarya" launched in 2000 contains complete data on water availability and use over a long period. The actual data on daily water discharges and levels of all hydraulic structures as well as on water volumes in the Naryn-Syrdarya cascade reservoirs form the basis of its information. Natural inflow to reservoirs has been recorded since 1911 and lateral inflow since 1948. Information input is carried out regularly during the day as online data arrives at the BWO's Central Control Office in Tashkent.

BWO uses the System for hydrometeorological information transmission of Central Asia using meteor burst communication on the terminal, through which hourly information on the most important basic gauging stations in the Syrdarya River basin is continuously represented in the Central Control Office in Tashkent. These data are irreplaceable in water-balance calculations, which form the basis for online water resource management and control of distribution among water users in neighbouring countries. For example, information received from the Uchkurgan gauging station enables the Control Office to track flow characteristics as they approach structures and allows timely adaptation of the Uchkurgan hydro scheme and the complex irrigation system on the BFC to operate under conditions of a rapidly changing regime of power-generating releases of the Naryn cascade. Water discharge data from the Kyzylkishlak gauging station, continuously transmitted by the system, allows for well-timed control of releases through the Kayrakkum hydropower station. Therefore, at the height of the growing season water is supplied according to assigned limits without detriment to users in Tajikistan, Uzbekistan and Kazakhstan. Thus, the existing potential enables BWO to solve problems of water resource management.

At the same time, we are on the verge of an acute water shortage and it is necessary to dwell on a question which should probably be solved in the near future in order to maintain economic security in all the countries of the region. It is a question of the operation mode for the Naryn-Syrdarya reservoir cascade. Initially, the main function of the Toktogul reservoir was to provide long-term regulation of the Naryn River flow in order to increase water availability for irrigated lands in an area of 9,18,000 ha, ensure growth of new lands by 400,000 ha in the Syrdarya River basin, and generate electric power amounting to 4.1 billion kW a year. Under this long-term regulation, the operation mode of the Toktogul reservoir was developed on the condition of guaranteed water supply to ensure sustainable crop yields. Its releases increase in low-water years and decrease in high-water years and amount in average water years to 9.43 km^3 or 75% of the annual volume in the growing season, and no more than 2.85 km^3 in the non-growing season. This makes it possible to keep as much water in the reservoir as possible and maintain high pressure. Thus, while satisfying irrigation needs, the largest energy output of installed capacities is provided not only for a particular year, but also for the whole hydrological period of alternating high and low water years. Moreover, a proper river regime adequate to a natural river hydrograph is arranged. Normal sanitary and environmental flows are maintained, and a land-reclamation regime of areas adjacent to the river is conserved.

It is also necessary to note that since 1974 the Toktogul reservoir has not had an opportunity to be filled, and its volume has not exceeded 5–6 km^3. When the long-term regulation regime is kept, 9 successive average or 2 high water years are required for its filling. During the high water years of 1987 and 1988 a maximum volume of 19.5 km^3 was achieved by August 1988, and the reservoir entered a period of total efficiency. In 1988 when the Toktogul reservoir first filled, a gradual deviation from the long-term regulation regime began. At the beginning, releases from the Toktogul reservoir increased during the non-growing season from 3.9 km^3 in the hydrological year 1989/1990 to 5.1 km^3 in 1991/1992. These changes did not strongly affect water supplies since they took place at the peak of high water availability in the year 1987/1988. The tendency to increase releases intensified further and the Toktogul reservoir is currently operated in a power-generating regime. As a result, the largest part of water releases and hydropower generation now take place in the winter period, and releases from the reservoir in the growing season are reduced to allow water accumulation. Such a change in the operating schedule has led to a complete deformation of river hydrograph, to floods in winter and to artificial water shortages in summer. The rejection of the long-term regulation regime leads to the following negative consequences:

- Acute water deficiency for irrigation
- Increased winter releases, which flood the areas and settlements adjacent to the channel and cause damage to the entire infrastructure in the downstream Syrdarya River since it is impossible to let the flow pass to the Aral Sea due to ice conditions
- Loss of environmental sustainability of water systems on the Syrdarya River due to under flooding and stoppage of return water in winter and channel drying in summer
- Emergence of the environmental problem in Aydar-Arnasai Lake system
- Possible deep drawdown of the Toktogul reservoir with the coming of sequence of low-water years

Considering the objective character of the reasons, which induced the Kyrgyz Republic to change the regime of the Toktogul reservoir, it was acknowledged that in order to prevent negative consequences it was necessary to compensate for water released from the Toktogul reservoir. At the same time the practice of compensation, according to the agreements of Kyrgyzstan with Kazakhstan and Uzbekistan established after 1995, to meet the needs of irrigated agriculture in the region during the growing season, does not concern the Toktogul reservoir regime in non-growing season which is made up based on electric power requirements of the Kyrgyz Republic. Under natural conditions the average inflow to the Toktogul reservoir amounts to 11–12 km^3 a year, and in recent years 15 km^3 a year is released from it, and it is clear without complicated calculations that during the last few years the reservoir is inevitably being emptied.

Foreseeing complications in the operation of the cascade, the states agreed and fixed in the Agreement of 17 March 1998, principles for compensation to provide rational use of water-energy resources in the Syrdarya River basin. This document confirms that deliveries and regimes of hydro schemes will be determined in yearly interstate agreements. In practice the difficulties with negotiations on the preparation of yearly agreements have not decreased but grown. The Toktogul reservoir still has a double load, which it is not able to carry. The situation of deep drawdown of the Toktogul reservoir arose in the winter of 1998 and in 2001. The current end of the non-growing season passed again for the regime of releases from the Toktogul reservoir without any limitations, as a result of which up to 8.5–9 km^3 will be released from it. The water level approached the dead volume mark by the beginning of the growing season, and only a high-water growing season can save the situation.

Deep drawdown of the Toktogul reservoir could be avoided through following a range, in which non-growing season releases should be controlled, and accordingly compensating the Kyrgyz Republic for water retained in the non-growing season. This will make it possible to gradually accumulate water supplies in the reservoir and eliminate water losses and damage caused by winter releases.

BWO "Syrdarya" has offered an intermediate version of a regime for the Toktogul reservoir: releases of 6.5 km^3 in the growing season, and releases of 4.9–5 km^3 in the non-growing season are recommended. At that, the volume of the Toktogul reservoir ensures the pressure needed for the installed capacity generation of 1,200 MW. Thus, we proposed a technical decision principally different from those being implemented in recent years, namely to compensate the Kyrgyz Republic for reducing non-growing season releases from the Toktogul reservoir, and releasing the water saved in this season on request during the growing season. We have brought our point of view to the attention of water authorities and Cabinets of Ministers of Central Asian states many times.

It is clear from the above that the operation mode for the Naryn-Syrdarya cascade approved by ICWC is fulfilled with substantial deviations due to the lack of an agreement on compensatory deliveries or non-fulfillment of the obligations stipulated in them. As a result, a discord in the water-management activities in the region has arisen. The operation modes of reservoirs are redesigned contrary to the schedules of ICWC, but it is not always possible to overcome the consequences of such failures. Water is lost and damage is done to water user sectors and the environment.

To solve this problem, in our opinion, water resource management and use should be completely concentrated in the hands of ICWC, which has all necessary rights and powers delegated by the countries of the region. Only ICWC should determine the implementation of regional water-management activities and a strategy for interstate water resource management in the Aral Sea basin.

IWRM Results in South Kyrgyzstan in Terms of Water Productivity Improvement

Barataly Turanovich Koshmatov

Many regions in the world are on the threshold of acute water shortage. Water use efficiency is low and the average efficiency of irrigation water use is 38%. The UN estimates irrigation system efficiency to increase by 42% (through advanced technologies and management practices) only by 2030. As a result of irrigation development, the mean annual flow withdrawal exceeds the environmentally allowable withdrawal in a number of river basins.

As is well known in our region, the largest per capita available water is in Kyrgyzstan and Tajikistan. It is 1.5–2 times lower in Kazakhstan, Uzbekistan, and Turkmenistan. In this context, let me quote water availability criteria as reported by the Canadian researcher Aly Shady: If per capita water consumption is higher than 1,700 m^3/year in a country, there is no water shortage; there are signs of water shortage if per capita consumption is between 1,000 and 1,700 m^3/year. If per capita consumption is less than 1,000 m^3/person, there is a water shortage with mostly unpredictable negative consequences. In the future, despite population growth and the fact that given countries possess different water quantities per capita, those criteria will define water shortage in any of the countries.

Water Challenges

During the Soviet period, irrigated agriculture was the basis of economic development in the republics of the Central Asian region, while the region itself was a raw materials-producing agrarian appendage of the Soviet Government. This provoked rapid growth of irrigated agriculture, which entailed new land development and irrigated area expanded by more than 1Mha between 1960 and 1980. The development of irrigation led to environmental deterioration all over the region as a result of the irretrievable

B.T. Koshmatov (✉)
Ministry of Agriculture, Water Resources and Processing Industry of the Kyrgyz Republic, Water Resources Department, 4a Toktonaliev Str., Bishkek 720055, Kyrgyz Republic
e-mail: pkti@elcat.kg

C. Madramootoo and V.A. Dukhovny (eds.), *Water and Food Security in Central Asia*, NATO Science for Peace and Security Series C: Environmental Security,
DOI 10.1007/978-90-481-9974-7_9,

diversions of huge quantities of water from rivers for irrigation and the massive application of chemicals in agricultural production. At that time the issue of improving water and land productivity was raised, especially on previously irrigated land. Though newly irrigated lands produced lower crop yields than lands previously irrigated, the former showed better water productivity through less irrigation water inputs per unit yield.

The climatic conditions and traditions in the Central Asian region made for a solicitous attitude of the local people to water and its use. However, with the development of large-scale irrigation, good traditions were left behind and a myth was formed about an inexhaustible abundance of water through the transfer of a share of Siberian river flow to Central Asia. The Aral Sea started to become catastrophically shallow, the Siberian river flow transfer project disappeared, and people in Prearalie again faced a challenge of water conservation and efficient water use.

Orientation towards potential water productivity may enable the region's countries to increase agricultural production almost twofold through a 10% reduction of water use. One example are WUFMAS data collected from monitoring factors contributing to crop yield on 220 control fields representative of Central Asian conditions. According to the data, annual irrigation water losses for "outlet to field – cotton rooting zone" average 51% throughout the region. The experience of countries producing up to 4 t of raw cotton per hectare under similar climatic conditions and at unit water inputs of no more than 5,000 m^3 per ha indicates a great potential for water conservation.

Many countries utilize waste water. In the Central Asian countries, waste water from industry and the municipal sector (6 $km^{3/}$year) is not used and creates a load on the environment. It is important to remember that water conservation is not only irrigation water conservation. It has been shown that saving water increases the productivity of irrigated agriculture. According to WUFMAS data on water and land productivity, the estimation of irrigation water use efficiency based on its unit inputs per unit of agricultural production shows that the mean cotton irrigation norm "gross-field" is 7,243 m^3 water/ha, including 2,039 m^3/ha (leaching and recharge irrigation) and 5,204 m^3/ha (growing-season irrigation). At mean cotton yields of 2.33 t/ha, the mean weighted irrigation water inputs per unit yield are 3,110 m^3/t and the water use productivity is 0.32 kg/m^3 on the field (indicators may vary from 1,600 to 10,340 m^3/t and 0.1 to 0.63 kg/m^3, respectively). For winter wheat, the mean weighted irrigation norm "gross-field" was 4575 m^3/ha. Average yields are 2.23 t/ha, the irrigation water inputs are 2080 m^3/t and the water use productivity is 0.49 kg/m^3 on the field (data vary from 180 to 5,750 m^3/t and from 0.17 to 5.65 kg/m^3, respectively).

Excessive water delivery to the field leads to lower land productivity since it leaches valuable nutrients from the soil which have to be replaced by costly mineral fertilizers, and increases groundwater levels and soil salinization under insufficient drainage. According to data from WUFMAS control fields, leaching losses of phosphorus and potassium are 65% and 50% of initial content. Moreover, soil salinization increases on average by 51% in 2 years.

The analysis of water losses due to organizational factors is very important. Such factors include mistakes in water distribution and water management particularly because of poor information, as well as excessive water losses in channels, and nonrecoverable outflows into closed sinks. As a result, huge volumes of water do not reach the Aral Sea. The regional and national experts should jointly identify those losses that amount to billions of cubic meters and develop mechanisms to avoid and prevent them.

Flow formation countries are ready to provide relevant information to all regional organizations dealing with flow forecasting and climate prediction. In order to ensure this, all the countries in the region should share financing of flow formation zone and of hydrometeorological activities.

Under market conditions, where economic mechanisms play a major role, water use efficiency and productivity and water conservation indicators at both national and regional levels will depend on charges for water resources and water use, while economic mechanisms of water conservation and of effective use will prevail over others. In fact, people do save something, which is not for free. As is well-known, Central Asian countries currently practice different approaches to irrigation water charges. In Kazakhstan, the payment for 1,000 m^3 of supplied water is 148.65 tenghe and the payment for every cubic meter of used surface water is 3.02 tiyn. In Kyrgyzstan, the payment for water supply is differentiated: 30 soms per 1,000 m^3 in the growing season and 10 soms per 1,000 m^3 in the non-growing season (I and IV quarters). In Tajikistan, 1 m^3 of supplied water costs 0.3 dirams. There are no irrigation water charges in Turkmenistan and Uzbekistan.

The experience of charging a fee for water use as applied in the Kyrgyz Republic is very important for saving water. According to analytical data of SIC ICWC, (prepared on the basis of databases processed and systematized in the SIC's Information Center under the projects "WARMIS" and "IWRM-Fergana"), after the introduction of water charges in1995, provinces in the Kyrgyz part of the Fergana Valley started to reduce irrigation water withdrawals. Before the introduction of water withdrawal charges (1986–1995), the annual water withdrawal was 3.7–4.7 Mm^3, while after charges were introduced, water withdrawals amounted to only 2.67–3.68 Mm^3 (excluding withdrawals in 1998 as the most humid year). The total withdrawal for irrigation in the three Kyrgyz provinces in the Fergana Valley over 5-year periods was 22,271 Mm^3 in 1986–1990, 19,655 Mm^3 in 1991–1995, and only 16,987 Mm^3 in 1996–2000 – the period of charged water use. Such reductions of water withdrawals took place under a constant irrigated area and minor changes in cropping patterns. Thus, we have every reason to say that the introduction of charged water use in the Kyrgyz Republic contributed to reduced irrigation water withdrawals and that the charged water use had a real effect on lowering water use and, to a certain extent, on improving water productivity.

Implementation in Kyrgyzstan of the policy aimed at introduction of water charges and establishment of Water User Associations contributed to reducing water withdrawals by 30%. Moreover, this takes place in a water-sufficient republic. However, taking into account a need to save water for the benefit of nature, Kyrgyzstan follows a water conservation policy, and implements integrated water resources management on the Aravan-Akbura canal, one of pilot canals within the IWRM-Fergana project. Within the framework of resource-conservation projects, Kyrgyzstan creates demonstration plots and pilot projects, where people are trained and educated in saving water and other resources. For example there are demonstration plots for projects such as WARMAP, GEF, IWRM-Fergana, as well as demonstration plots for energy-saving under the SPECA project.

Food Challenges

To address food challenges, the Central Asian states follow a grain independence policy. In this context grain areas, mainly wheat, were increased considerably at the expense of cotton, forage, vegetable and cucumber areas. Moreover, livestock productivity decreased as a result of lack of an efficient forage base, poor quality forage and feeding processes and low assimilability of forage. Production of vegetables, fruits and cucumbers completely meet local demand and are exported. However, actual export of such goods does not correspond to the capacity of the region due to export limitations. As a result, agricultural producers lose their income.

Food imports are reduced in terms of volume and assortment. The policies followed by the states are aimed at further reduction of imports and increasing food exports and do not imply regional specialization. A considerable gap between the growth rates of food supply and population leads to an aggravation of food problems. Access of population to adequate nutrition will depend on the efficiency of agricultural production and on national economic development in general.

Water is an important factor under such conditions. One major indicator of nutrition is the quantity of calories. According to medical data, calorie consumption per capita should be 2,700–3,200 kcal. This will require adequate agricultural production to meet food demand. In order to achieve this, measures should be taken in the near future to strengthen agriculture, expand crop areas and, undoubtedly, all these actions will be connected with a need to develop and use water resources.

Aiming at further improvement of irrigation techniques and crop yields, future irrigation development will likely be based on the following positions. First, irrigated agriculture needs to be intensified in order to meet people's demand for food. Second, water use in irrigated agriculture should be evaluated on the basis of future advanced water-conservation technologies. At the same time, it is necessary to consider well-proven traditional water-conservation techniques. Thus, one of the ways to reduce norms in surface irrigation is a system of field-protection plants, which leads to saving 15–25% of irrigation water. Third, in the future we should rely on abrupt increases in crop yields. For example Mexican irrigated wheat has become widespread in the tropical zone. This wheat can yield 10 t per ha. It served as the basis for the "green revolution" in India and solved food problems in other countries. Obviously, the world food problem cannot be solved using technical measures only. Biology plays a very big, perhaps, leading role in this respect. Fourth, by using biotechnologies and advanced irrigation techniques (for example, drip and other ways of subsoil irrigation), we may be able to achieve the expected results, including provision of water resources. Fulfilling the task in terms of the two important aspects of water resources, quantity of water diversion from sources and water quality, one can meet mankind's demand for water in order to solve the food problem.

Food growth will be ensured mainly through an increase in crop yields. Moreover, irrigation will decrease through advanced irrigation techniques and selection. The total irretrievable water input will increase slightly. This indicator should become stabilized and the main effect will be reflected in the reduction of water input per 1 t of production. Therefore, the main strategic line of water sector development to solve water quantity and quality problems is the reduction of water input per unit of production. This should be addressed in all branches of the water sector, particularly in irrigated agriculture.

The development of integration processes between the Republics is of importance. Taking into account the long-term bread-grain price forecasts, the problem related to lack of grains for food in the region can be solved through cooperation by producing grain in Kazakhstan. It is well-known that growing wheat on irrigated land under current yields is not cost-effective. Therefore, the argument that wheat self-provision is economically reasonable is not cogent. Dry wheat grown in the Kazakh steppes produces better results in terms of yields and profit. By following a moderate policy of grain self-provision, based on the development of interstate cooperation and a possibility to cover a bread-grain deficit from Kazakhstan, it is possible to:

- Expand areas for forage crops, mainly alfalfa.
- Renew crop rotation, use alfalfa as a nitrogen-fixing crop for soil fertility restoration.
- Increase production of fodder grain.
- Raise efficiency of forage base for livestock.
- Improve livestock productivity.

The regional cooperation should be based on the economic benefit from production and address such areas as meat and milk production in Kazakhstan and Kyrgyzstan, grazing development and sugar beet production in Kyrgyzstan, earlier vegetable production in Turkmenistan and Uzbekistan, and fruit and cucurbits production in Tajikistan, Uzbekistan, and Turkmenistan. Hotbed farming and using film technologies for early vegetable production, and canned and dried fruit production will contribute to increased export capacities. Intra-regional cooperation will lead to reduced food imports, allowing small exports in the region, based on the cost-effectiveness of producing certain agricultural products in some zones. It is clear that no one country can agree on a policy which entails considerable dependence on food imports until this country achieves long-term economic and military security.

According to human development forecasts, the urban population is expected to grow, with accompanying water re-distribution. The water use priority would be provision of needed quantities of water for urban zones, industry and services. The agricultural sector will rely more on the production of high-value crops. As world experience shows, despite the potential production benefit due to advanced technology, inevitably, water will be re-distributed from agriculture to more valuable uses in water-scarce regions. The governments of Israel, Cyprus, and Malta successfully re-employed their people in other activities, including industry, commerce, and tourism. Agriculture is limited mainly to high-value export crops and food is not produced in those countries. This is a strategy of replacing irrigated agriculture by food imports compensated through urban and commerce development (the so called import of "virtual water"). In our region, one should recognize a need for and the value of "virtual water" (water used for non-agricultural purposes, for food import) for regional food provision.

Since the collapse of the USSR, the transition to new economic relations has not been easy in the economic sectors of the Central Asian countries. Agriculture faced problems such as a lack of equipment, fertilizers, and chemicals, as well as a deficit between the output and input prices.

As of January 1, 2007, the area of the Kyrgyz Republic is 19,995.1 thousand ha. There are 10,766.4 thousand ha of agricultural land of which: 9,176.1 thousand ha are under pasture; 1,283.7 thousand ha are arable, and the rest falls to other land categories. The total irrigated area is 1,020.6 thousand ha of which 866.3 thousand ha is irrigated arable land. There are diverse soil and hydrogeological conditions even within the boundaries of one province in Kyrgyzstan. Depending on those parameters, irrigation conditions differ and irrigation efficiency depends largely on soil, water management and irrigation technology.

Effect of Rehabilitating WUAs on Crop Yield

The "On-farm irrigation" Project plans to improve crop yield through stable and reliable water distribution on about 1,20,000 ha of irrigated area. To track project results and the impact on agricultural production, a special database was developed for information collected from all WUAs. Using this database, the impact of rehabilitating WUAs, particularly crop yields before and after rehabilitation, is analyzed and assessed. Unfortunately, it is impossible to estimate increases in the crop yield due only to additional water quantities as this increase is a sum of many factors, such as optimal agrotechnical dates, high-quality seeds, fertilizers, etc.

By December 31, 2007, the rehabilitation was completed in 51 WUAs. According to the project implementation plan, an assessment of rehabilitation impact on yields should be made at the end of the second year after work completion. Yield analysis, made in 14 WUAs, where one or three growing seasons had passed since rehabilitation, showed variable results. Winter wheat yield was increased by 8–19% in half the WUAs; tobacco yield increased by 2–7% in all WUAs; yield of cotton increased by 6–18% in 2/3 WUAs and corn yield increased by 15–20% in half WUAs.

The rehabilitation impact was assessed by ORP and an invited independent company. Results show that WUAs subjected to rehabilitation supported higher crop yields than WUAs not supported by the project. The survey of the independent company shows that 86% of farmers indicated improved reliability of water supply, 82% of farmers noted reduced water losses, 88% of farmers received water in time (observance of water delivery schedules), and 91% of farmers indicated a more equitable water distribution.

Analysis of Irrigated Areas

For each planning zone defined by a certain river reach and irrigated command area inside the Republic and for the Republic in general within a basin area, the following indicators and factors should be determined, analyzed and evaluated:

- Potential water and land productivities – based on available data of best practices, especially in dry years.
- Unit water consumption under minimum water inputs into organic production, using common technical approaches.
- Causes of low production (due to land and water-related factors) and measures to overcome this, with definition of priority measures to be undertaken.
- Salt and water balances of planning zones, using historical data; measures to bring these parameters into the range leading to environmentally sustainable processes (minimum salt exchange between rivers and irrigated areas and between aeration zone and groundwater, with gradual reduction of salt storage in the aeration zone and in the planning zone in general).
- Measures to maximize use of return waters and utilize them directly in places close to their formation.
- Measures to use wastewater, groundwater, and water from all local sources that are not currently utilizable.
- Measures to reduce water losses caused by organizational factors in all chains of water-distribution hierarchy.
- Unproductive water losses in all chains of irrigation system, first on irrigated field; estimation of these losses will help the choice of less capital-intensive measures for water conservation.
- Reduction of return water discharge into rivers and water bodies and improvement of water quality as a result of water conservation measures.

The main factors of low irrigation water use productivity are:

- Unstable irrigation water availability in canals.
- Improperly selected schemes and parameters of irrigation technology.
- Low quality of field leveling and preparatory agro-technical measures.

The main indicators of low water and land use efficiency are:

- Large losses to infiltration; large losses to discharge from irrigated fields.

- Non-observance of times to carry out separate technological operations and the quality of their performance.
- Low doses of fertilization or their absence; insufficient measures for control of weeds, diseases and pests.

The main ways for improving water productivity and water conservation in the region are:

- Introducing water charges in irrigated agriculture through establishing incentive step tariffs and fine sanctions per cubic meter of water used over established norms and so on.
- Developing common technical approaches to rigid standardization of water consumption based on more precise norms mainly designed for meeting the minimum biological needs of plants.
- Creating a system of pilot water conservation projects as first-priority entities of demonstrative water use.
- Introducing water rotation and other organizational measures aimed at control of water losses in field and its non-productive use (short furrows, irrigation with concentrated stream through furrow, thorough maintenance of field leveling and so on).
- Introducing advanced irrigation techniques and technology.
- Installing impervious membranes on canals.
- Integrated and partial reconstruction (modernization) of irrigation systems.

Considering that most losses occur in the field and in water distribution among newly privatized farms, the establishment of Water User Associations is the most important mechanism for regulation of water use and conservation at this level along with charged water use.

Assessment of the Efficiency of "IWRM - Fergana" Project's Impact in the Republic of Uzbekistan

Sergei Alexeyevich Nerozin

Introduction

Water productivity is an indicator characterizing the "coverage" of unit water delivered for irrigation by crop yields in physical and cost terms and is described by the following formula:

Water productivity (kg/m^3) = Crop yield (kg)/Quantity of water used (m^3)

Application of this indicator per hectare allows water management to be assessed adequately at hierarchical levels, analyzes current tendencies in water use, and searches for measures to reduce water losses in irrigate agriculture.

Dynamics of Water Resources Use in Fergana and Andizhan Provinces of the Republic of Uzbekistan

Information received from the Fergana and Andizhan Basin Administrations for Irrigation Systems (BAIS), as well as from the provincial agricultural authorities served as a basis for the estimation of irrigation water productivity per rayon.

Tables 1 and 2 illustrate the dynamics of efficiency of irrigation water use by the rayons of Fergana and Andizhan provinces for 2005–2008 and indicate a certain tendency for improving water use both within the project zone and in non-project rayons. However, at present, it is quite difficult to adequately assess the degree of such impact using statistical data at the rayon level, since there exist multiple factors (soil, institutional, water-management, technological, etc.), as well as spatial characteristics that impact water and land productivities.

S.A. Nerozin (✉)
SIC ICWC, Regional expert, B.11, Karasu-4, Tashkent 700187,
Republic of Uzbekistan
e-mail: info@icwc-aral.uz

C. Madramootoo and V.A. Dukhovny (eds.), *Water and Food Security in Central Asia*, NATO Science for Peace and Security Series C: Environmental Security,
DOI 10.1007/978-90-481-9974-7_10,

Table 1 Dynamics of irrigation water productivity (kg/m^3) by rayon of Fergana province (2005–2008)

	Cotton				Spiked cereals			
Rayon	2005	2006	2007	2008	2005	2006	2007	2008
Besharik	0.397	0.312	0.353	0.442	0.464	0.495	0.580	0.667
Bagdad	0.326	0.262	0.312	0.447	0.520	0.578	0.662	0.582
Buvaydy	0.473	0.388	0.296	0.379	0.494	0.530	0.605	0.550
Dangary	0.343	0.290	0.446	0.437	0.505	0.520	0.543	0.650
Yazyavan	0.400	0.303	0.363	0.488	0.420	0.434	0.526	0.401
Kuva	0.320	0.212	0.276	0.307	0.532	0.565	0.763	0.569
Oltiarik	0.322	0.282	0.430	0.418	0.487	0.562	0.821	0.664
Ahunbabayev	0.325	0.247	0.392	0.411	0.360	0.328	0.580	0.465
Rishtan	0.268	0.310	0.451	0.521	0.464	0.533	0.670	0.511
Sokh	–	–	–	–	0.958	1.323	2.200	1.278
Tashkak	0.312	0.333	0.295	0.314	0.489	0.523	0.652	0.511
Uzbekistan	0.247	0.210	0.246	0.353	0.443	0.530	0.504	0.485
Uchkuprik	0.350	0.313	0.447	0.451	0.457	0.501	0.484	0.487
Fergana	0.184	0.178	0.201	0.216	0.515	0.529	0.575	0.589
Furkat	0.363	0.291	0.368	0.464	0.462	0.541	0.567	0.737

Project rayons

Table 2 Dynamics of irrigation water productivity (kg/m^3) by rayon of Andizhan province (2005–2008)

	Cotton				Spiked cereals			
Rayon	2005	2006	2007	2008	2005	2006	2007	2008
Altinkul	0.543	0.481	0.822	0.653	0.746	0.947	0.714	1.049
Andizhan	0.365	0.256	0.292	0.376	0.873	0.866	0.966	1.100
Asaka	0.301	0.185	0.349	0.397	0.750	0.554	0.743	0.915
Balikchy	0.632	0.428	0.666	0.719	0.856	0.866	0.865	0.994
Boz	0.396	0.282	0.459	0.456	0.612	0.704	0.673	0.939
Bulakbashy	0.386	0.281	0.382	0.425	0.970	0.731	0.794	1.219
Djalalkuduk	0.283	0.332	0.339	0.369	0.949	0.929	1.058	1.236
Izbaskan	0.734	0.392	0.512	0.553	0.783	0.783	0.962	1.278
Kurgntepa	0.349	0.367	0.373	0.495	0.913	0.906	0.960	1.263
Markhamat	0.320	0.266	0.370	0.444	0.839	0.715	0.769	1.179
Pakhtaabad	0.531	0.316	0.398	0.462	0.855	0.894	0.800	1.059
Ulugnar	0.328	0.181	0.256	0.276	0.510	0.725	0.462	0.643
Khadjaabad	0.340	0.253	0.250	0.384	0.659	0.707	0.727	1.137
Shakhrikhansai	0.544	0.338	0.560	0.542	0.815	0.684	0.789	1.023

Table 3 Irrigation water productivity (kg/m^3) by WUA of Andizhan province (2005–2008)

		Cotton				Spiked cereals			
Rayon	WUA	2005	2006	2007	2008	2005	2006	2007	2008
Project impact zone									
Khadjaabad	Khodjaobkash	0.365	0.365	0.425	0.561	0.706	0.766	0.710	0.950
	Garagura	0.358	0.353	0.424	0.439	0.684	0.740	0.684	0.712
Bulakbashy	S. Kasimov	0.402	0.409	0.452	0.456	0.985	1.043	0.897	0.966
	A. Giyasov	0.392	0.400	0.438	0.451	1.011	1.004	0.911	0.946
Markhamat	Tomchi kuly	0.358	0.315	0.481	0.625	0.890	0.896	0.970	0.907
	T. Mirzayev	0.352	0.305	0.447	0.462	0.875	0.905	1.022	1.072
	Mashal	0.354	0.312	0.492	0.499	0.884	0.912	0.983	0.997
Average for WUA		**0.369**	**0.351**	**0.451**	**0.499**	**0.862**	**0.895**	**0.882**	**0.936**
Beyond the project impact zone									
Khadjaabad	A. Nabiyeva	0.322	0.249	0.244	0.250	0.624	0.691	0.702	0.715
	Ittifok	0.338	0.252	0.237	0.235	0.652	0.683	0.700	0.720
Bulakbashy	Aybek	0.358	0.273	0.353	0.368	0.884	0.714	0.794	0.805
	Ulugbek	0.365	0.271	0.362	0.365	0.879	0.725	0.790	0.817
Markhamat	M. Tojibayev	0.300	0.260	0.325	0.322	0.825	0.696	0.592	0.601
	Pakhtakor	0.308	0.267	0.336	0.331	0.832	0.708	0.604	0.591
	Bobur	0.319	0.262	0.347	0.349	0.732	0.682	0.594	0.584
Average for WUA		**0.330**	**0.262**	**0.315**	**0.317**	**0.776**	**0.700**	**0.682**	**0.690**

Efficiency of Irrigation Water Use in WUAs (Within and Beyond the Project Impact Zone)

The main aim of Water User Associations (WUA) is managing water resources in the command area and ensuring sustainable, equitable and equal water distribution among water users. At present, 40 and 20 WUAs organized on the hydrographic basis are functioning in Fergana and Andizhan provinces, respectively. To assess the performance of WUAs, statistical information on individual associations located within and beyond the project impact zone has been collected and processed. In 2005–2008, an increase in irrigation water productivity was observed while growing strategic crops in most WUAs of Fergana and Andizhan provinces, located in the project impact zone (Tables 3 and 4).

Table 5 shows a clear upward tendency of irrigation water productivity in the project impact zone for cotton and spiked cereals for 2005–2008. Water productivity (average for WUAs) slightly decreases beyond the project impact zone. This means that water and land resources were used less efficiently there.

Table 4 Irrigation water productivity (kg/m^3) by WUA of Fergana province (2005–2008)

		Cotton				Spiked cereals			
Rayon	WUA	2005	2006	2007	2008	2005	2006	2007	2008
Project impact zone									
Fergana	Isfayram	0.34	0.37	0.39	0.42	0.64	0.70	0.93	0.91
	Tursunaly	0.61	0.60	0.62	0.61	0.99	0.96	1.13	1.08
	Sattorov	0.35	0.24	0.36	0.39	0.67	0.70	0.75	0.70
	Kuchkorboy	0.28	0.26	0.28	0.30	0.56	0.57	0.61	0.56
Tashlak	Khonobod	0.62	0.61	0.65	0.71	1.43	1.12	1.13	1.10
	Varzak	0.60	0.58	0.63	0.76	0.93	1.11	0.91	0.92
	Arab Shermat	0.54	0.57	0.58	0.66	0.81	0.95	0.94	0.96
	Fargkumar	0.60	0.62	0.67	0.72	0.90	0.97	1.01	1.06
Kuva	Omad	0.57	0.59	0.63	0.65	1.38	1.42	1.49	1.53
	Tolmozor	0.44	0.55	0.51	0.57	0.90	0.91	1.03	1.11
	Zilol	0.45	0.44	0.49	0.50	0.77	0.81	0.83	0.92
	Guliston	0.39	0.46	0.50	0.51	0.75	0.81	0.85	0.90
Average for WUA		**0.48**	**0.48**	**0.53**	**0.57**	**0.89**	**0.92**	**0.97**	**0.98**
Beyond the project impact zone									
Язяванский	Khonobod	0.48	0.48	0.45	0.50	0.74	0.74	0.70	0.73
	Ok – Oltin	0.48	0.49	0.42	0.44	0.64	0.67	0.63	0.60
	Z. Ganiyev	0.29	0.30	0.27	0.33	0.40	0.36	0.35	0.34
	Yangibuston	0.24	0.22	0.22	0.28	0.31	0.33	0.26	0.33
Average for WUA		**0.37**	**0.37**	**0.34**	**0.39**	**0.52**	**0.53**	**0.49**	**0.50**

Table 5 Growth rate of irrigation water productivity (%) within and beyond the project impact zone (average for WUA) 2005–2008

	Cotton				Spiked cereals			
Crop	2005	2006	2007	2008	2005	2006	2007	2008
Fergana province								
Project impact zone	100	100	110.4	118.8	100	103.4	109.0	110.1
Beyond the project impact zone	100	100	91.9	105.4	100	101.9	94.2	96.2
Andizhan province								
Project impact zone	100	95.1	122.2	135.2	100	103.8	102.3	108.6
Beyond the project impact zone	100	79.4	95.5	96.1	100	90.1	87.9	88.9

Assessment of Efficiency of Irrigation Water Use in Rayon Demonstration Areas (2005–2008)

Aiming at disseminating advanced technologies and accumulated project experience, in 2005 pilot rayon demonstration areas (farms) were organized in the Fergana and Andizhan provinces of the Republic of Uzbekistan. Advanced technological methods and measures for improving land and water productivity were implemented and demonstrated for surrounding farms. This allowed all

Table 6 Dynamics of irrigation water use efficiency (kg/m3) in demonstration area of Andizhan province for 2005–2008

		Cotton				Spiked cereals			
Rayon	Farm (demonstration area)	2005	2006	2007	2008	2005	2006	2007	2008
Andizhan	Orzu tashab	0.34	0.35	0.49	0.61	0.78	0.81	0.64	0.83
Asaka	Bozorboshi	0.42	0.46	0.61	0.62	0.83	0.55	0.87	0.79
Balikchy	"Omad"	0.43	0.44	0.57	0.57	0.88	0.94	1.08	1.38
Buz	Khojumurod	0.45	0.47	0.56	0.58	0.63	0.68	0.68	0.72
Bulakbashy	Talibjan	0.71	0.79	0.73	0.72	1.01	1.2	1	1.43
Jalakuduk	Sardor	0.43	0.52	0.66	0.59	0.73	1.01	0.81	1.27
Izboskan	Pakhtakor	0.71	0.74	0.72	0.61	0.97	1.12	1.29	1.05
Ulugnor	Bakhori	0.3	0.32	0.41	0.54	0.46	0.51	0.6	0.57
Kurgantepa	Berdiboy	0.39	0.67	0.51	0.64	0.98	1.29	1.21	0.99
Markhamat	Khosildoy	0.26	0.28	0.52	0.6	1.1	1.62	0.75	0.91
Altinkul	Bakht	0.51	0.53	0.57	0.61	0.99	1.03	1.05	1.48
Pakhtaabad	Abdurakhim	0.39	0.4	0.52	0.52	0.77	1.09	0.94	0.83
Khujaabad	Yer va el	0.39	0.5	0.46	0.56	0.98	0.97	1.09	0.88
Shakhrikhan	Dilshoda	0.46	0.54	0.68	0.58	0.9	0.76	0.92	1.07
Average for farms (demonstration areas)		**0.44**	**0.5**	**0.58**	**0.61**	**0.86**	**0.97**	**0.92**	**0.99**

Table 7 Dynamics of irrigation water productivity (kg/m3) in demonstration areas of Fergana province

			Cotton				Spiked cereals			
No	Rayon	Farm (demonstration area)	2005	2006	2007	2008	2005	2006	2007	2008
1	Besharik	Nosir ota	0.71	0.76	0.77	0.79	0.87	0.92	1.08	1.2
2	Furkat	Dildorabonu	0.45	0.48	0.54	0.62	0.8	0.83	0.97	0.95
3	Tashlak	Ergashboy	0.41	0.45	0.48	0.53	0.86	0.9	1.15	1.16
4	Kuva 1	Turdaly	0.77	0.87	0.8	0.76	1.34	1.55	1.63	1.74
5	Kuva 2	Ismoilov	0.57	0.66	0.67	0.62	1.62	1.72	1.8	1.76
6	Rishtan	Davronbek	0.54	0.61	0.59	0.63	1.26	1.5	1.66	1.53
7	Yazyavan 1	Koratepa	0.59	0.59	0.63	0.66	0.82	0.95	0.86	0.9
8	Yazyavan 2	Dustlik	0.7	0.86	0.75	0.79	1.69	0.99	1.27	1.18
9	Akhunbabayev	Khojalkhon	0.4	0.41	0.53	0.57	1.08	1.17	1.31	1.76
10	Buvayda	Sarbon	0.41	0.46	0.52	0.55	0.95	0.98	1.23	1.3
11	Altiarik	Ortikov	0.4	0.47	0.49	0.58	0.63	0.75	0.88	0.93
12	Uchkuprik	Arabkurgon	0.46	1.03	0.55	0.58	0.91	0.98	1.42	1.14
13	Dangara	Nasivaly ota	0.59	0.7	0.78	0.63	1.34	1.47	1.02	1.07
Average for farm (demonstration area)			**0.54**	**0.64**	**0.62**	**0.64**	**1.09**	**1.13**	**1.25**	**1.28**

agrotechnical actions that are conducted during growing and non-growing seasons to be corrected. Soil, climatic and institutional conditions of the rayons were taken into account and farmer's agricultural knowledge and the training process were improved. Dynamics of the efficiency of irrigation water use in rayon demonstration areas for 2005–2008 indicate a considerable increase in these indicators in the majority of rayons (Tables 6 and 7).

The rayon water use productivity indicators, averaged for Andizhan province (Table 6), increased by 38% for cotton by 2008, compared with the base year 2005 (from 0.44 to 0.61 kg/m^3). These indicators also increased for cereals from 0.86 kg/m^3 (2005) to 0.99 kg/m^3 (2008). Growth of water productivity is also observed in Fergana province (Table 7) and for cotton amounted to a 37% improvement and for grain a 17% improvement. As a whole, the results obtained for the rayon demonstration areas should be considered good and this allows recommending reasonably wide dissemination of the established agricultural practices among farms.

Irrigation Water Productivity in Farms Located near Rayon Demonstration Areas in Fergana Province

The chosen rayon demonstration areas are ordinary farms, owners of which agree to be a base for demonstrating advanced technologies and disseminating best practices in agricultural production. Within the framework of the project, the farmers of the demonstration areas together with province executives actively participate in training farmers on neighboring farms, and disseminating booklets and technological bulletins. Such training covered 214 farms in Fergana and 240 farms in Andizhan provinces.

The comparative assessment of irrigation water productivity in farms located near the rayon demonstration areas (generalized data) shows a clear tendency for improving water resources use (Table 8). The results indicate the usefulness and efficiency of activities on disseminating water and agricultural knowledge and skills among the farmers. Obviously, it is possible to achieve higher indicators for crop yields and water saving provided that institutional and financial-economic problems concerning farming are solved (liberalization of prices for production inputs, increasing purchase prices for agricultural products, tax deductions, easier access to credit etc.).

Conclusions and Proposals

1. Despite relative water sufficiency in Fergana and Andizhan provinces, the uniformity of water use for strategic crops (cotton, spiked cereals) is quite unstable, indicating certain problems in agricultural water management and use.

Table 8 Comparative assessment of irrigation water use efficiency (kg/m^3) in farms located near the rayon demonstration areas

	Cotton				Winter wheat			
Province	2005	2006	2007	2008	2005	2006	2007	2008
Fergana	0.418	0.440	0.450	0.446	0.785	0.843	0.890	0.890
Andizhan	0.427	0.457	0.533	0.520	0.643	0.701	0.799	0.797

2. The mean (over the project period 2002–2008) irrigation water productivity was 0.29 kg/m^3 for cotton and 0.66 kg/m^3 for cereals in Fergana province. In Andizhan province irrigation water productivity was substantially higher: 0.35 kg/m^3 for cotton and 0.94 kg/m^3 for cereals. This indicates much better irrigation water and agricultural production management in Andizhan province.
3. At present, it is quite difficult to assess adequately the degree of project impact at rayon level since there exist multiple factors (soil, institutional, water-management, technological, etc.), as well as spatial characteristics that affect water and land productivities.
4. The summarized data on WUAs over 2005–2008 demonstrate a clear upward tendency in irrigation water productivity in the project impact zone for cotton and cereals. Water productivity decreased slightly in cotton and grain fields (average for WUAs) beyond the project impact zone. It should be noted that many WUAs, despite their legitimate status and certain rights, are not functioning efficiently: more than 50% of WUA's directors have no specialized water-related education; associations lack professional staff; salaries are handed out with delays that depend on collection of funds from water users; provincial and rayon departments of the Uzbekistan's Ministry for Agriculture and Water Resources make small contributions to development and support of WUAs.
5. For sustainable and effective functioning of WUAs, it is proposed (as an experiment in two pilot WUAs) to revise payment by farmers of irrigation service fees and estimate amount of contributions from the farmers to WUA using the following principle:

   - For strategic crops (cotton, spiked cereals), setting a fixed charge for WUA services at 6–7 USD per ha
   - For other free marketed crops, setting a charge in the form of 6% deductions from the income generated on an area of 1 ha (orchards, vineyards, vegetables, cucurbits)

6. In order to disseminate the improved technologies and accumulated project experience, in 2005 in Fergana and Andizhan provinces, the pilot rayon areas were established to apply and demonstrate modern technological approaches and methods of water and land productivity improvement for adjacent farms. This made it possible to adjust all agrotechnical operations undertaken in growing and non-growing seasons, taking into account soil, climatic, institutional, and technical conditions in any rayon and to ease training of farmers and improve their agricultural knowledge. Dynamics of irrigation water productivity and efficiency in rayon areas during 2005–2008, as well as agro-economic data indicate a substantial increase of the indicator values both in most rayon areas and in given provinces as a whole. It is significant that economic indicators in the areas are 20% higher and more on the average in terms of profit as compared to ordinary farms in the Republic. The establishment of extension services and respective encouragement of their staff would increase farmer income as a whole, raising their paying capacities and, hence, improving financial conditions of WUAs.

7. An upward trend of irrigation water productivity over 2005–2008 is observed from the summarized data on farms located near rayon areas in Fergana and Andizhan provinces. The achieved results indicate the usefulness and efficiency of activities on dissemination of water-agricultural knowledge and skills among farmers. Obviously, it is possible to achieve higher indicators for crop yields and water saving provided that institutional and financial-economic problems are solved (liberalization of prices for production inputs, increasing purchasing price for agricultural products, tax deductions, obtaining credit, etc.).
8. It is necessary to increase the involvement of water-management organizations (especially at the rayon level) in the implementation of IWRM-Fergana Project's results, dissemination of accumulated experience among WUAs and farms, development of extension services and encouraging water users to participate in water management.

Issues of Capacity Building and Training in IWRM Implementation for Achieving Socio-Economic Stability Through Higher Productivities of Water and Land Use in the Region

Pulatkhon Umarov

The capacity building program for the joint management of transboundary water resources, according to the fundamental provisions of the regional water strategy, and developed from existing scientific, educational and financial capacities for the Aral Sea basin countries (Fundamental provisions of the regional water strategy for the Aral Sea basin and Tashkent 1996) includes:

- Institutional improvement of the regional organizations network and development of a legal framework for shared transboundary water management
- Implementation of integrated water resources management (IWRM)
- Development of a training system
- Creation of regional and national information systems
- Implementation of automated water control systems and communication networks, including SCADA systems
- Development of modeling tools and decision support systems

Training plays a key role in capacity building and is related directly to all other parts of capacity building that, in turn, influence the development and improvement of the training system itself (Fig. 1). An impact on any one capacity building factor automatically touches all others.

The major and basic achievement in institutional improvement and legal framework development for the shared water management in the Aral Sea basin, was that water leaders from the Central Asian states (Republic of Kazakhstan, Kyrgyz Republic, Republic of Tajikistan, Turkmenistan and Republic of Uzbekistan) signed in February 1992 an Agreement on the establishment of the Interstate Commission for Water Coordination (ICWC) with its regional executive bodies: BWO "Amudarya," BWO "Syrdarya," SIC ICWC and ICWC Secretariat. This

P. Umarov (✉)
Scientific Information Center of Interstate Commission for Water Coordination (SIC ICWC), Training Centre of ICWC, B.11 Karasu-4, Tashkent 700187, Republic of Uzbekistan
e-mail: umarovp@icwc-aral.uz

C. Madramootoo and V.A. Dukhovny (eds.), *Water and Food Security in Central Asia*, NATO Science for Peace and Security Series C: Environmental Security,
DOI 10.1007/978-90-481-9974-7_11, © Springer Science+Business Media B.V. 2011

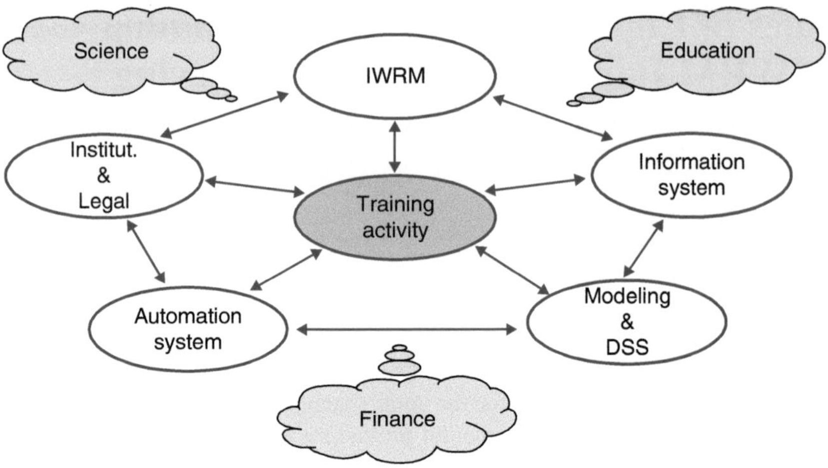

Fig. 1 Capacity building development

event paved the way to capacity building for cooperation in the field of shared water resources management in the Aral Sea basin.

In its present-day interpretation, IWRM is a contemporary of ICWC, since it became widely and officially recognized only after the Dublin Conference in 1992, and is now the most popular management system. At the same time, it is said that "new is well forgotten old"! Indeed, if we look to the history of water sector development, prototypes of this system at the beginning of the twentieth century are easily found:

- Spanish water confederations
- French water agencies
- North American and Canadian irrigation districts

In the past, the foundation for this system in our country was laid in the following studies:

- Prof. Rizenkampf, G.K. 1930. Need for an integrated approach to water sector development
- Schemes for integrated water resources use in river basins (Volga, Don, etc.)
- Management within hydrographic boundaries (ZERDOLVODKHOZ in the Zerafshan River basin, 1926; Amudarya Delta and Derivation Canal Water Management Organization – UPRADIK in the Amudarya River basin, 1927)
- Integrated development of large massifs (Hunger steppe 1956; Karshi steppe 1964; etc.)
- General schemes for integrated water resources use and protection in the Aral Sea basin 1956

Looking back on the historical water sector experience in our country and comparing it with recent achievements of our foreign colleagues, we can proudly state that our science and practice were undoubtedly at a sufficiently high level. Nevertheless, there were also some drawbacks such as underestimation of environmental interests and opportunities for public participation in water resources management.

The initiator of theoretical development and practical implementation of IWRM in the region is Prof. V.A. Dukhovny, according to whom "IWRM is a system of management based on consideration of all kinds of water resources (surface, ground and return) within hydrographic units that harmonizes the interests of different sectors and water use hierarchy levels, involves all stakeholders in decision-making, facilitates effective use of water, land and other resources to meet environmental and human water requirements in a sustainable manner" (Dukhovny and Sokolov 2006).

To acknowledge the practical advantages of IWRM and demonstrate the feasibility of outcomes expected from its implementation under production conditions, SIC ICWC with the support of the Swiss Agency for Development and Cooperation (SDC) initiated a regional project titled "Integrated Water Resources Management in Fergana Valley," involving the water management organizations of Kyrgyzstan, Tajikistan and Uzbekistan. The results are evident and it is generally recognized that the following have been achieved in the project area:

- Stable water supply, uniform and equitable distribution of water resources under considerable reduction of non-productive losses in water supply systems
- Democratization of management through involvement of all stakeholders in management as well as in maintenance and development of systems
- Improvement of water and land resource use productivity
- Solution to environmental problems, including land reclamation
- Improvement of human well-being

The popularization of IWRM in the region was preceded by gathering and disseminating knowledge about the current tendencies in world water sector development as well as by holding awareness-raising workshops demonstrating the experience of such countries as Canada, France and Israel in IWRM implementation. The experience of the 1970s and 1980s, when SANIIRI at the Ministry of Water Resources of USSR held regular scientific-practical workshops for advanced training of water specialists in support of developing countries under programs of the UN Economic Commissions for Africa, Asia and the Pacific, Latin America and the Caribbean and others, was of use in this activity. By the way, a regional branch of the All-Union Institute for Advanced Training of Water Specialists of Caucasus, Central Asia and Kazakhstan was then founded in SANIIRI.

Even after the collapse of the USSR and the break in existing relations, under conditions of economic difficulties during the transition period and sharp decline of scientific-technical potential in the sector in all the countries of the region, SANIIRI (now in SIC ICWC status) continued to strengthen sector workforce capacity through the training of young specialists and development of advanced methods for informatics, management, economics and law. While maintaining and developing relations with the International Commission on Irrigation and Drainage (ICID), the UN Economic Commissions, UNESCO, FAO, CIDA, USAID, MASHAV, NATO and others, experience and information exchanges were initiated, including holding of regional workshops in Tashkent and abroad.

The combination of such experience in regional and international scientific-technical cooperation was also conducive to a considerable degree to the formation of ICWC, and further development of the Aral Sea Basin Programs, TACIS and GEF projects. Project Terms of Reference stipulated a training component and familiarization trips to raise the awareness of sectoral leaders of the need for water management system reconstruction according to IWRM principles. Leader training helped to achieve required political support, provided a needed public understanding of oncoming water crises on a planned basis, and realized the urgent need to implement IWRM through transforming it to a national concept.

At the same time, realizing the need for ongoing advanced training of water specialists, it was decided at the 21st ICWC meeting on 24 October 1998 to establish a Training Centre and ask the Canadian International Development Agency (CIDA) for financial support. Thus, a 5-year Project "Water Resources Management in Central Asia" started in 2000 with the support of CIDA and in partnership with McGill University and Mount Royal College (Canada). The Training Centre began its activity at an inception workshop with the participation of ICWC leaders and executive bodies according to the plan for advanced training of water specialists of higher and medium levels approved by ICWC. This workshop determined a training policy and a strategy aimed at understanding the foundations of water problems not only by water specialists, but also by leaders of society, government and non-governmental organizations, and other stakeholders involved in pilot projects at each level of the water hierarchy.

Advanced vocational training, as one important line of capacity building, is a continuous and dynamic process (Fig. 2). This process comprises training needs assessment (TNA), training strategy (TS), training activity (TA), training efficiency assessment upon completion of workshop program or feedback assessment (FBA) and after a certain period of time, when trainees have departed and started working or follow up assessment (FUA), as well as fine tuning of the existing training strategy (FTETS), based on repeated training needs assessment (TNA) in a new cycle.

At the initial stage, training covered senior staff members of respective ministries and departments, employees of provincial and basin administrations, and then medium-level specialists of water management organizations.

The fundamental principle of the strategy for training was to propagandize and agitate for the urgency of IWRM ideas for the region and transform it into a national

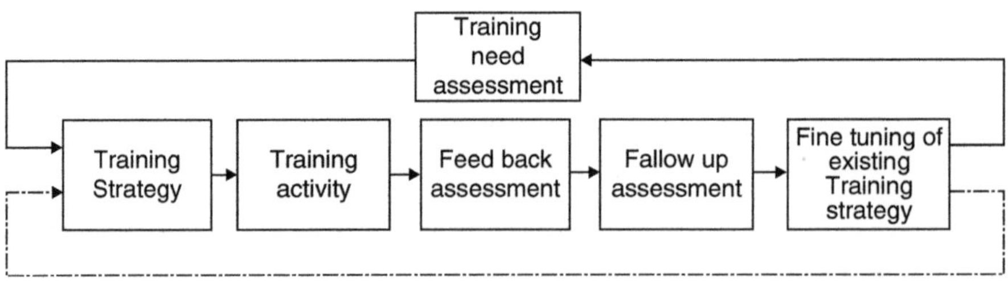

Fig. 2 Common framework of training process development

program for actions aimed at water sector reform. The active participation of water sector leaders from ICWC member countries in training provided all-round support for development of social mobilization of water users and all stakeholders, training in national IWRM plans and adoption of them by the national governments. Such a strategy for training was also aimed at wider public recognition of the need for cost-effective solutions and non-governmental funds for water sector improvement through development of organizational and public forms of participation. All that contributed to the promotion and recognition of IWRM in official legal and regulatory documents of the countries in the region (Dukhovny and Sokolov 2006).

The major distinctive feature of such a training system is that it is based on the results of research activities carried out under interstate programs and different regional projects such as "IWRM-Fergana"; Strategic planning of IWRM; IWRM in downstream river reaches; IWRM in river deltas; Drainage problems in Central Asia; Water and education; Regional IWRM model in twinned river basins "RIVERT-WIN"; Strategic planning of regional cooperation; Development of a modeling tool based on relationships of water resources, socio-economic development, and nature protection for training of and use by decision makers; Modeling the optimal operation of Naryn-Syrdarya reservoir cascade; "CAREWIB" Information System; Automation of SFC head structures and so on. Results of these activities are regularly included in lecture materials and presentations used in training. The cooperation and partnership with colleagues from leading international universities and institutes, including McGill University and Mount Royal in Canada, IHE–UNESCO and ILRI in Holland, Montpellier and Cemagref in France, Bonn and Stuttgart Universities in Germany and

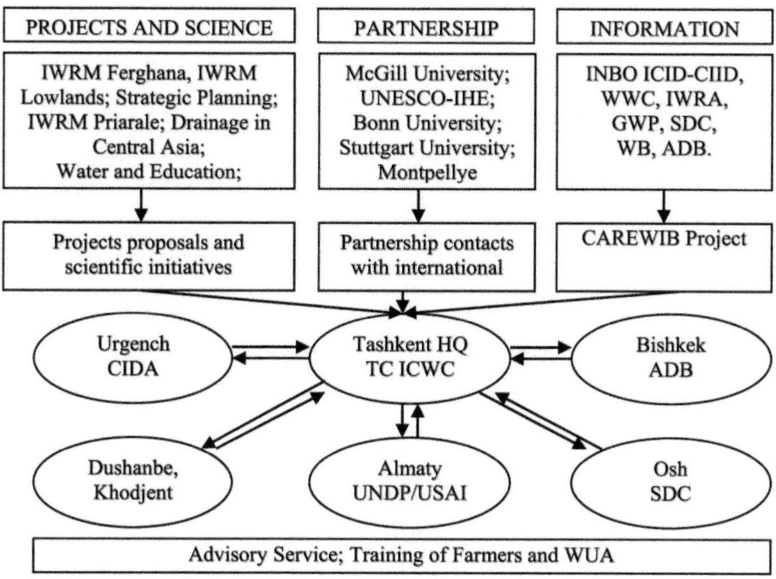

Fig. 3 Information and knowledge dissemination and exchange within and beyond the training network

so on are also conducive to regularly updating presentation materials, widening the themes of workshops and improving the procedure for training with inclusion of situation modeling elements (Fig. 3). The regular information exchange, which was established thanks to the membership of SIC ICWC and election of its Director for the governance membership of such organizations as ICID, WWC, INBO, IWRA, GWP, and the support of such donor organizations as CIDA, WB, SDC, ADB, USAID and others, serves to fulfill these tasks. It is also appropriate to mention here that the establishment of the GWP Technical Committee for Central Asia and Caucasus was also initiated by the Director of SIC ICWC, and the first organizational workshop with participation of the Secretariat of GWP Headquarters (Stockholm, Sweden) was held in the ICWC Training Centre (Umarov 2006).

The other distinctive feature of the training is the improvement of its effectiveness through decentralization and the establishment of Training Centre branches in Osh town with the support of SDC (for trainees from provinces of Fergana Valley within the borders of the Kyrgyz Republic, Republic of Tajikistan and Republic of Uzbekistan), in Urgench city with the support of CIDA (for trainees from pre-Amudarya provinces and districts of Turkmenistan, Karakalpakstan and Republic of Uzbekistan), in Almaty city with the support of USAID and in Bishkek city with the support of ADB. It is also planned to establish sub-regional branches in Dushanbe city (for trainees in South Tajikistan and Surkhandarya, Kashkadarya provinces in the south of Uzbekistan) and Kzyl-Orda town (for trainees from rice-growing areas in the south of Kazakhstan and north of Uzbekistan).

In addition to the Training Centre branch in Osh where workshops for the provincial and district water management organization level and canal and WUA level are held, local training centers were set up in Fergana, Andizhan and Khodjent towns within the framework of the "IWRM-Fergana" Project. Furthermore, an experimental training center was established on the basis of the WUA model "Akbarabad" in Kuva District Center in Fergana province for training farmers, representatives from WUA, dekhkan farms, makhalla and village committees. Results of activities at lower levels show that achieving goals to improve water use and productivity requires dissemination of this experience through extension services. Therefore, further training development requires establishing special training points in demonstration fields where mobilizer-trainers, trained at higher levels of the training pyramid, can train farmers not only in handling methods for water accounting, use and distribution, but also in all issues related to improvement of land and water productivity as a part of a consulting assistance.

The head office of the Training Centre and its branches interact as follows (Table 1):

- Coordination of regional training activity, development of new training courses, training of employees at higher and medium levels and preparation of program, lecture, methodical and technical materials for branches are done at the head office in Tashkent;
- Training of employees at lower levels of water management organizations is carried out by branches through moderators participating in the preparation of lectures and

Table 1 Main structure and framework for organization of training by the ICWC TC

Responsible hierarchy	Target groups	Type of training and facilities	Main thematic directions
High level (TC headquarters)	Heads of department and divisions of Ministries of agriculture and water management; environmental protection; economy and finance; foreign affairs and justice	Training of Trainers, Training the trainees Interactive discussions and in future e-learning	▪ Regional cooperation on Transboundary water resources management ▪ International and national water law and policy ▪ IWRM on basin level and Water allocation planning ▪ Strategy and approaches to Advanced irrigated agriculture
Middle level (TC headquarters and National branches of TC)	Heads of Basin and systems of water management organizations, NGOs dealing with water management; representatives of agriculture and water education, design and research institutions	Training the trainees. Interactive discussions and in future e-learning	▪ Basic principles of IWRM ▪ Advanced irrigated agriculture ▪ Water resources information system development as a tool of regional cooperation ▪ National water low and linkage with international ▪ Ecosystem and hydro-informatics approaches ▪ Financial and economic sustainability of water sector ▪ Public awareness and participation
Low level (National branches of TC and local focal points)	WUAs, farmers, extension services representatives; representatives of agriculture and water education, design and research institutions	Training the trainees Practical training	▪ Integrated water resources management ▪ Water use and accounting at the farm level ▪ Water and land productivity ▪ Irrigation and drainage ▪ Extension services

delivered materials, using the network of demonstration plots under pilot projects, database and integrated information systems of the regional centre of SIC ICWC.

The no less important characteristic is the thematic scope of workshops and involvement of representatives from related sectors in training. The themes of workshops are represented by the following courses:

- Integrated water resources management
- cooperation on transboundary rivers
- Water law and policy
- Improved irrigated agriculture

Along with water specialists of Central Asian countries, representatives from ministries and departments of the environment and energy and from non-governmental organizations were invited to the first two courses, while representatives from the ministries of foreign affairs and justice of each country involved in preparation of interstate agreements and national legislative acts in the field of water management and environmental protection were also invited to the workshops on water law and policy held with participation of specialists from Great Britain (Dundee University) and Israel (Israeli Center for Negotiation and Mediation).

The training system provides for monthly workshops lasting 6–7 days for 20–30 people of equal representation from the five countries in the head office. Between times there is a series of 3–4 day workshops for the same number of people in the branch offices with participation of trainer-moderators from head office and lecturers from among trained representatives of local organizations.

The adopted interactive training format based on the preliminary dissemination of lecture and supplementary materials, exchange of views and discussions between trainees on problems and current challenges of the water sector, using experienced moderators (often ICWC members), promotes mutual understanding, common ground, openness and trust. In essence, every training workshop turns into a round table of representatives from different countries and different sectors and facilitates consensus-building in the region at inter-sectoral and interstate levels through brainstorming stimulated by moderators and lecture theme. Moreover, a protocol and joint recommendations are prepared based on the results of every workshop, which are subsequently sent to ICWC members for further dissemination and for permanent improvement of the system.

Another important achievement of such regional training is the creation of conditions for establishing friendly and direct contact between specialists from different countries and sectors related to water resources management problems during joint lectures and leisure time. It is especially important to consider the fact that the trainees, who are young specialists today, may become CEOs of local and national authorities, production departments and even sectors tomorrow, and each of them may become a decision maker in the future.

About 5,000 specialists have been trained in the ICWC Training Centre on experience and principles of IWRM as a whole, fundamentals for water legislation at national and international levels, improvement of irrigated agriculture, intensification of public participation, gender aspects of IWRM and so on (Table 2).

To maintain contact with its trainees, the Training Centre has created a special database, to which current information on each trainee is entered. This database is updated every year with consideration of all changes in labor activity of its trainees, part of which is involved in its activities in the branch offices. Thus, specialists, who have been trained at the head office of the Training Centre in

Table 2 Number and breakdown of trained specialists per year

Countries/ Year	Kazakhstan	Kyrgyzstan	Tajikistan	Turkmenistan	Uzbekistan	Other countries	Total
2000	9	11	11	6	12	–	49
2001	63	45	55	27	99	7	296
2002	176	171	160	128	204	5	844
2003	141	189	164	130	259	1	884
2004	173	145	144	80	427	1	970
2005	149	107	126	56	392	9	839
2006	96	85	84	28	286	15	594
2007	67	59	71	15	226	–	438
2008	19	65	45	4	220	18	371
Total	893	877	860	474	2125	56	5285

Tashkent, subsequently train trainees from lower level water management organizations, WUAs and farmers, using program modules, lecture and methodical materials as well as visual aids developed in the head office. At the same time, an effective many-stage system forms where each trainee transfers his/her knowledge to colleagues at his/her and lower levels ("domino" principle). Information on activities of the Training Centre and its current and planned workshops is continually updated on a special website.

As an example of qualitative assessment of the training system, we consider the general course on integrated water resources management, which most trainees from specialists of higher and medium levels as well as representatives of lower level water users or WUAs have passed. Having them understand the importance and necessity of reforms in the sector through implementation of advanced principles for water resources management provided favorable conditions for promoting not only the above-mentioned "IWRM-Fergana" Project, but also the project for transboundary water resources management in Chu-Talas basin, that broadly involved the public and considered the interests of all stakeholders.

All of that, in turn, resulted in widening the circle of associates and dissemination of an understanding of IWRM as a way for future survival of the planet on the one hand, and strengthened the partnership of water specialists to solve water problems on the other hand. Opportunities for implementation of integrated water resources management in downstream reaches in the Aral Sea basin based on the same principles are currently being considered, and in Uzbekistan it is planned to carry out similar measures in the Zeravshan River basin.

The improved practical understanding and involvement of specialists from water management organizations and water users in IWRM is conducive to a gradual representation of these ideas in such important legal documents as Water Codes of Kazakhstan, Kyrgyzstan and Tajikistan, Laws on WUA of Kazakhstan, Kyrgyzstan and Tajikistan, Decree of the President of Uzbekistan "On major ways for deepening reforms in agriculture" and Decision of the Cabinet of Ministers of Uzbekistan "On improved organization of water sector management."

The same positive shifts can be observed in terms of the results of workshops on cooperation in transboundary rivers and water law and policy, which brought together stakeholders involved in the work of inter-sectoral conciliation commissions on interstate agreements. The experience of workshops and joint protocols adopted by the participants undoubtedly prove that the Aral Sea basin countries can successfully and effectively solve the problems of water supply and electric power generation only through effective mutually beneficial cooperation. Such cooperation should be based on the principles of hydro-solidarity, respect and consideration of interests of all the countries, and minimizing damage to irrigation and power generation as well as to the environment, using the previously built capacity under current market conditions.

Separatist tendencies slow the work of promoting agreements on information exchange and the organizational framework of regional bodies signed earlier by ICWC members and which previously dominated interstate water relations. The current widening of the circle of associates in other ministries and departments contributes to a better understanding of the need for consolidation based on cooperation and willingness to revive the functioning of inter-sectoral conciliation commissions in each of the countries.

The same can be noted regarding workshops on improving irrigated agriculture, which explain that in the context of water deficiency, the only way to reduce the difference between water supply and demand in the region is by water demand management using water conservation methods. In the provinces there is a growing understanding that it is possible to achieve potential water productivity and enable an increase in agricultural production almost 2-fold while reducing water use up to 10%. This in turn determines the willingness of states to direct investments into activities for water conservation through the creation of a network of demonstration plots providing extension services to farmers, WUAs and water management organizations. Holding workshops dedicated to problems of water conservation and water and land productivity improvement enabled the identification of prospects for the involvement of women in solving these socially important issues and a special program “Water – gender – agricultural productivity improvement” was prepared.

To raise public awareness, a series of workshops were held, inviting representatives from non-governmental organizations and the mass media. At these workshops, the participants studied the current problems of water sector and the environment, including peculiarities of reform and democratization of water sector management. The need for involving the public in the movement for implementation of IWRM and water conservation, popularizing it as a way for survival under conditions of a growing water deficit is also explained.

To form the outlook of a new generation of future water users oriented to a respectful attitude to water, a series of workshops under the “Water and Education” Project were held with the support of OSCE and in partnership with the Ministry of Popular Schooling of Uzbekistan. These workshops were for teachers of institutes for advanced training and teachers of primary and secondary schools who are familiarized with IWRM elements in order to integrate it into school educational programs.

The main condition for ensuring efficiency of the created training system is its permanency and continuity. This is important not only because of personnel renewal throughout the hierarchy of water sector management and the migration of rural population to cities that is occurring, but also because the acquired knowledge and experience that must be transferred to water users is continually updated as new solutions from trial procedures appear. Therefore, IWRM training at the level of administration and departments chiefs of ministries (committees, departments, main directorates) of water resources should be held in Tashkent, Urgench, Almaty and Bishkek at least twice a year, before and after vegetation period.

Furthermore, for the level of leaders of provincial water management organizations, basin administrations of irrigation systems as well as their deputies and managers of divisions for water use, representatives of local authorities and specialists of WUAs, workshops on IWRM should also be held at least two times a year in each province of the countries in the region. At these workshops, attention should focus on the following issues:

- Organizational and legal aspects in improving the system for implementation of WUAs, Canal Water Users Unions (CWUU), Water User Groups (WUG)
- Improvement of the system for water use, accounting and reporting in water use at the level of WUA, WUG, CWUU
- Organization and carrying out measures for land reclamation as well as repair and renewal measures in fields and at irrigation and drainage systems

For steady capacity building, IWRM and training should be developed dynamically along with the water information system network that is being developed and permanently upgraded. This system network includes the modeling tool and decision support system, the creation of a network of automated systems for water infrastructure management in the region, the institutional improvement and creation of a legal framework for joint management of transboundary water resources. These activities were begun in 1996–1997, when the WARMIS regional information system and a database on management of water and land resources with GIS sub-base was created within the framework of the WARMAP project under the European Union TACIS Program. This system is continually developed and improved, and populated with information separately for each country on the following blocks: economy, surface water, groundwater, land, climate, industry, administrative and territorial division, environment in the Aral Sea and Priaralie. At present, this activity continues with the financial support of SDC within the framework of the "Central Asia Regional Water Information Base (CAREWIB)" Project representing CAWater-Info portal with information on water and environmental problems in Central Asia and the regional information system. This system includes the knowledge base, database with GIS, information of hydromet services on flow formation, information of BWOs on water allocation, and analytical block with a set of modules and models for assessment of the current and predicted 1-year and multi-year water situation. The project is aimed at building and strengthening capacity of water management organizations by attracting them to the permanently functioning network of correspondents actively interacting with each other at regional, basin and national levels, through the formation of national information systems closely linked to regional, basin and national databases. To ensure sustainable functioning of CAREWIB, initial systematic training is organized for specialists of national information systems on a common methodical basis for building an information system with transfer of software and maintenance of available databases on countries and provinces as well as for maintenance and improvement of national databases to provide regular information exchange and population of the regional database with relevant data. Such work on development of interactions between specialists of water management organizations at all levels and provision of free access to information for all stakeholders will undoubtedly facilitate capacity building and successful implementation of IWRM in the region.

Activities for equipping interstate head intake structures with Supervisory Control and Data Acquisition (SCADA) systems were started in terms of BWO "Syrdarya" structures with the financial support of USAID and are being continued with support of SDC on main canals of the IWRM-Fergana project. The introduction of this system with electronic means for information receipt, storage and transmission makes it possible not only to prevent uncontrolled river water withdrawals, but also improves water supply accuracy (to 2% instead of 10% before) and reduces non-productive water losses. Furthermore, the creation of a common system for automated transmission of technological information provides all stakeholders with free access to daily updated actual data on water discharge and level at all hydro structures, facilitating trust in joint water resources management. For building capacity of specialists of operational organizations in charge of this system, special training workshops are held in the ICWC Training Centre, its branch offices and at SCADA sites. It is planned to begin the application of a SCADA system at BWO "Amudarya" soon with the support of ADB.

The creation of a legal framework for joint management of transboundary water resources was also started within the framework of the WARMAP project and four draft agreements that specify organizational setup, information exchange, water use regulation and environmental protection were developed. To continue the improvement of a legal framework for regional cooperation, working groups authorized by national governments, consisting of representatives from different concerned sectors, were established and their activities organized on permanent basis. At the same time, measures for improving national water legislation were carried out through inserting in it the principles of IWRM as the main tool for improving water productivity at all levels of water hierarchy. At present, this activity is supported by ADB and implemented within the framework of the Project "Strategic planning of regional cooperation (RETA)" aimed at development of a program for future legal and organizational measures for capacity building of interstate water cooperation in the Aral Sea basin. Based on the results of this work, draft agreements on information exchange, on enhancement of ICWC status, on water and energy resources use in the Syrdarya River, and other documents on procedures and rules for river basin management were prepared and discussed at regional workshops.

The inclusion of results achieved by all of these training projects and methodical materials of the Training Centre maintains a permanent cycle of improvement of the training system and continuity of the capacity building process.

In further activity on capacity building, it is necessary to broaden training themes, involve specialists from related sectors such as energy, environment, water supply, hydromet services in work on improving management and rational use of water resources. It is also necessary to foresee special workshops in order to prepare a basis for development of public participation in integrated water resources management and establishment of a network of non-governmental organizations interacting with water management organizations in each country. There is a need to organize training on project management in the fields of water management and environmental protection, economic reforms, agricultural production, including fish production in irrigation systems. Furthermore, the scope of activity should be expanded with training programs on international organizations and financial institutions.

Thus, the created system of capacity building for the water sector helps specialists better gain world experience, outlines paths of movement "from vision to action," defines priorities, raises his/her work to new levels of modern computerization, informatics, use of the internet resources and globalization. Along with enhancement of professional skills and familiarization with recent achievements in the field of water and land resources management, irrigation and drainage, environmental protection, the training system helps to strengthen the spirit of cooperation between countries in the region in the field of use and management of water resources and develops common approaches at the level of specialists and decision-makers. Therefore, being a conductor of all recent achievements in the field of water management and land reclamation, the ICWC Training Centre has also become a political institution for capacity building for regional cooperation of Central Asian states on the way to practical implementation of IWRM at system and basin levels in connection with national priorities and regional interests, while keeping a balance between national requirements and regional limitations.

References

Dukhovny, V.A., Sokolov, V.I. (2006). Integrated water resources management. Experience and lessons of Central Asia, Tashkent.

Fundamental provisions of the regional water strategy for the Aral Sea basin, Tashkent, 1996.

Umarov, P.D. (2006). Developing human and institutional capacity in transboundary water management through water resources training network, Tashkent.

Scenarios of Future Development in the Aral Sea Basin

Anatoly Sorokin

Present Status of the Water Sector

The water utilization scheme (WUS) in the Aral Sea basin is a combination of inter-related elements of water-management systems of the Syrdarya River basin and the Amudarya River basin. In functional terms, WUS is characterized by the processes of water generation, consumption and production, as well as by interactions between humans and the environment (see Fig. 1). The main objectives of WUS include: meeting the water demands of society and national economies; and conserving ecosystems for future generations. The primary water consumer is irrigated agriculture and the primary water user is hydropower.

The surface water resources in the Aral Sea basin comprise renewable river waters and return waters. The total average annual river water volume is 116.5 km^3/y in the Aral Sea basin, including 37.2 km^3 in the Syrdarya basin and 79.5 km^3 in the Amudarya basin. Approximately 32.5 km^3 of return waters (collector and drainage waters from irrigation and wastewater from industry) are formed in the Aral Sea basin, of which 13.4 km^3 is in the Syrdarya basin and 19.1 km^3 is in the Amudarya basin. About 16.8 km^3 of return water is discharged into rivers (9.2 km^3 – Syrdarya basin; 7.6 km^3 – Amudarya basin), 10.9 km^3 into natural sinks, and 4.8 km^3 are re-used. Usable water resources in the Aral Sea basin (natural river run-off and return flow into rivers) are estimated on average at 133 km^3, minus channel losses of 117–121 km^3. In the Aral Sea basin, the total water diversion was 60.6 km^3 in 1960 and achieved a maximum of 120.7 km^3 in 1980. It decreased to 104 km^3 by 2000 and stabilized at 105 km^3 in 2008. Water diversion per irrigated hectare was at a maximum (16.9 thousand m^3/ha) in 1970 and a minimum (11.9 thousand m^3/ha) in 2000. The maximum diversion per capita, 4.73 thousand m^3/y, was observed in 1970, while the minimum was 2.53 thousand m^3/y in 2000. Since 1992, interstate water allocation limits have been practiced in the basin. The targeted tendency is the reduction of water

A. Sorokin (✉)
Scientific Information Centre of Interstate Commission for Water Coordination (SIC ICWC), Tashkent
e-mail: info@icwc-aral.uz

C. Madramootoo and V.A. Dukhovny (eds.), *Water and Food Security in Central Asia*, NATO Science for Peace and Security Series C: Environmental Security,
DOI 10.1007/978-90-481-9974-7_12,

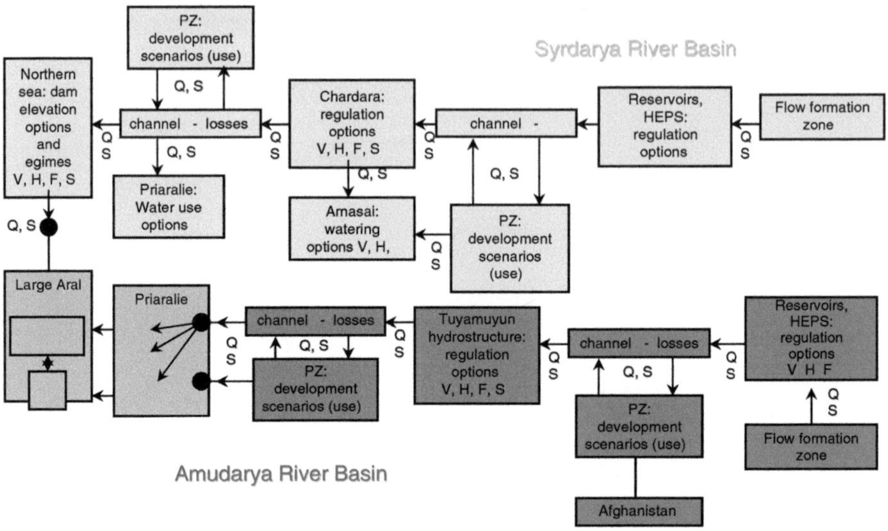

Fig. 1 General scheme of water management for the Aral Sea basin

diversions (as a result of the water conservation policy adopted by ICWC). At the same time, changes in irrigated agriculture, i.e. reduction of cotton areas and increase of areas under grain and other crops also decreased diversions.

In the Aral Sea basin, there are more than 60 reservoirs, with a total useful capacity of 46.5 km^3, including 20.2 km^3 in the Amudarya basin and 26.3 km^3 in the Syrdarya basin. Forty-five HEPS, with a total capacity of 34.5 GWt, are operational. Hydropower accounts for more than 27% of the total energy consumption in the basin. This indicator varies substantially among the countries: most hydropower is produced in Tajikistan (about 98% of the total national electricity generation) and in Kyrgyzstan (approximately 91%), while the lowest hydropower production is in Turkmenistan (1%). Because of constructed reservoirs, the degree of flow control is 0.94 along the Syrdarya River (natural runoff controlled fully) and 0.78 along the Amudarya River (available reserves for further control).

Since 1992, demands for energy and operation regimes of reservoir hydrosystems with HEPS have changed and resulted in reduced water availability for irrigated agriculture in the middle and lower reaches of Amudarya and Syrdarya in some dry years. Meanwhile, the guarantee of annual necessary electricity generation by large interstate HEPS, such as Toktogul and Nurek, is still low due to an inefficient regime of water accumulation and drawdown of the reservoirs that fails to maintain required heads at HEPS. For example, since 1992, summer releases along the Syrdarya River have depended heavily on supplies of electricity, fuel, and gas from Kazakhstan and Uzbekistan to Kyrgyzstan by way of barter. Water shortages during the growing season and forced discharges into Arnasay sink during the non-growing season have occurred. As a result, not only irrigated agriculture but the whole ecosystem of the Syrdarya basin has become dependent on those regimes. Table 1 and Figures 2 and 3 show water releases from the Toktogul reservoir for two characteristic periods – before and after 1992.

Table 1 Inflow into and releases from Toktogul reservoir over characteristic periods (km^3)

	1985–1991		1992–2006	
	Non-growing	Growing	Non-growing	Growing
Inflow	2.77	9.29	3.20	10.49
Release	3.53	7.93	8.13	5.45

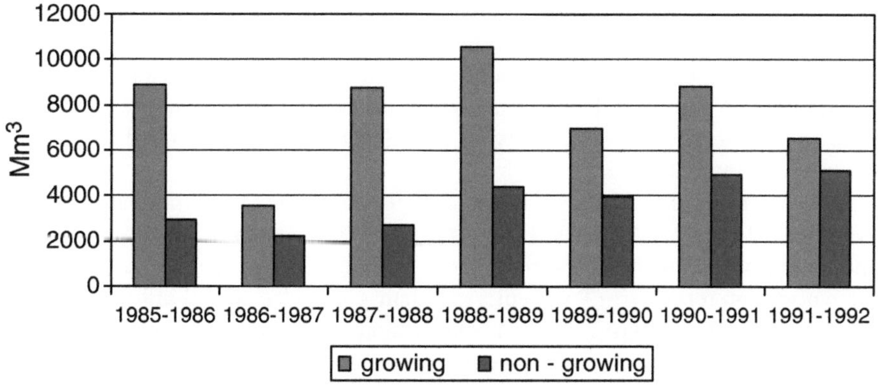

Fig. 2 Water releases from the Toktogul reservoir from 1985 to 1991

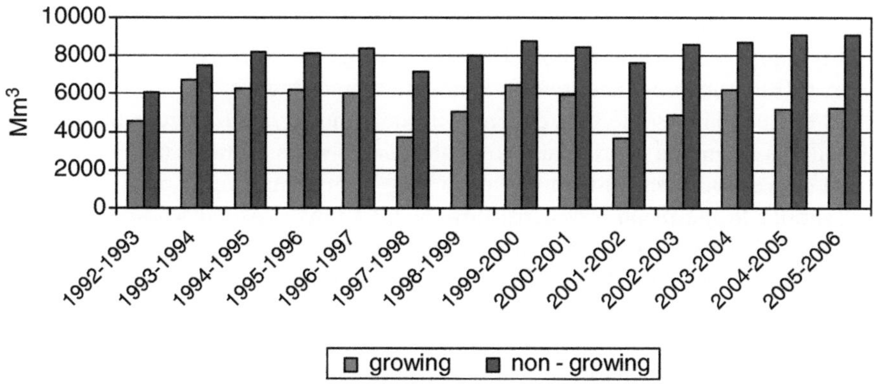

Fig. 3 Water releases from the Toktogul reservoir from 1992 to 2006

Understanding the environmental requirements, especially since the beginning of ICWC activity, has promoted an increase of inflow to ecosystems of the Amudarya and Syrdarya deltas and the Aral Sea over the last 10–15 years. However, water supply to the ecosystems is often connected with availability of excessive runoff along the rivers due to energy-related releases from reservoirs. Existing aquatic ecosystems (Arnasay Lake system and others) require certain annual releases to stabilize levels and the water-salt balance, and, at the same time, there is a risk of excessive releases or undersupply of water to ecosystems due to high runoff fluctuations caused by natural factors (river fluctuations) and inefficient flow control by reservoirs.

National interests of riparian countries dictate a need for correction of the operation regimes of the interstate reservoir hydrosystems with HEPS in order to improve energy efficiency and reduce irrigation water deficits and prevent eventual conflicts between the states. Today, regulating reservoirs in the basin is not enough. The solution is seen at the interstate level through the development and adoption by the states of rules of flow regulation along the Syrdarya and Amudarya Rivers and their main tributaries. In order to increase guaranteed water yields in the basin, it is necessary to implement long-term regulation and put into operation new reservoirs against the background of water and energy conservation.

The need for developing the flow regulation rules is justified not only to improve management but also to solve difficulties in climatic and hydrological forecasts that determine both expected flow fluctuations in the rivers that flow into reservoirs and necessary generation of electricity at HEPS. This, in turn, depends on energy system requirements as a whole.

Run-off from the Syrdarya River and its tributaries over the last 17 years averaged 41.6 km^3/y. This is 3.4 km^3 (or 8%) higher than the mean long-term and annual values over 1950–1990. The same situation is observed for inflows to three upstream reservoirs (Toktogul, Andizhan, Charvak): annual inflow for the last 17 years averaged 24 km^3, that is 1.7 km^3 higher than the mean long-term annual flow over 1950–1990. If we compare mean annual flows in the Syrdarya River over 17 years with the mean long-term flow of 37.6 km^3 over the whole observation period 1911–2007, the increase in flow will be even more (10%) for 17 years. Frequency of low-water years in the Syrdarya basin over the 17 years did not become higher as compared to 1950–1990; however, high-water years (25% probability and lower) became more frequent (1.4 times), and extremely high-water years (10% probability and lower), almost two times higher.

The situation is somewhat different in the Amudarya basin. Run-off from the Amudarya River and its tributaries over the last 17 years averages 69.2 km^3. This is 1 km^3 lower (only 1.5%) than the mean long-term annual value over 1950–1990, but similar to the mean long-term flow of 69.3 km^3 over the whole observation period (1911–2007). Low- and high-water years in the Amudarya basin became more frequent over the last 17 years compared to 1950–1990. The frequency of low-water years (75% probability and higher) increased 1.3 times, while that of high-water years (25% probability and lower) became higher by 1.2 times, and for extremely high-water years (10% probability and lower), by 2.5 times. The "depth" of extremely low-water years increased 1.5 times (i.e. deviation of the mean flow in low-water years from that over the period). Thus, in recent years, not only floods (for all the rivers) and low-water periods (for Amudarya) became more frequent but the amplitude of deviation from the mean values has increased as well.

Future Basin Development Scenarios

Each state in the basin has its own development concepts, strategic interests and priorities that do not always agree with other states, and also has a specific vision of regional conflict resolution. However, there exist key problems for all the states.

These are related to current and future water allocations between the states in the basin and to river run-off regulation. Here it is important to estimate impartially future available water resources and water demands. Vagueness of information in this area increases the risk of investments in the water sector and undermines national and regional initiatives on the improvement of water management. Regional scenarios for the future should imply coordination, provision of incentives and support of national measures. At the same time, national scenarios should be checked against regional restrictions using such planning tools that can weight national estimations and indicators and avoid mutually exclusive management options. In this case only it can be expected that national results can be aggregated to the regional level. The efficient tool for developing future scenarios and strategies is an integrated planning method, which implies setting interstate and intersectoral links, common objectives, search of common spheres of interests, based on trade-offs and compensations, and search of consensus.

The main factors impacting future water development in the Aral Sea basin are as follows:

- Dynamics of future runoff hydrographs and of unit crop water use as determined by climatic changes
- Dynamics of future water requirements for users (economic sectors and the nature) and of return waters as determined by national water sector development concepts and demographic trends
- Dynamics of future country energy needs, export and import of fuel and energy resources, putting new reservoirs and HEPS into operation, operation regimes of interstate reservoir hydrosystems with HEPS as determined by national energy concepts
- Dynamics of losses of regulating capacities due to siltation, which depends on future water and sediment runoffs and reservoir operation regimes

Thus, the assessment of the future situation in the basins can be made by scenarios that estimate risks from increased water use, climate changes, reservoir capacity losses, new interstate reservoirs and HEPS, etc. All probable scenarios of basin development (based on the above factors) can be combined into two integrated scenarios (schemes) that are named conditionally as

- Pessimistic scenarios – "A", based on business as usual in national water and energy sector development and on existing interstate agreements, including provisions and factors increasing the risks of development
- Optimistic scenarios – "B", suggesting better indicators of national water and energy sector development and improved interstate agreements

These schemes include only feasible objectives and targets, excluding imbalances caused by excess of water demands over available water (attempts to integrate national strategies into a regional one). Scenario "A" envisages maintaining the existing principles of reservoir operation (before and after new reservoirs are put into operation), while scenario "B" foresees changes in operation regimes, according to developed flow regulation rules that are based on long-term regulation and compensation principles.

Dukhovny et al. 2008 give justification of a need for adaptation to climate change in CA. In the authors opinions, as a first response to more frequent extreme hydrological phenomena, we should pay heightened attention to more intensive long-term regulation and to ensuring guaranteed water storage in reservoirs of interstate importance. The European Convention's Commission on Climate and Water at their meeting in Bonn suggested more frightening figures on the Amudarya basin: the expected decrease of water resources due to diminishing glaciers and glacier component could be 30%! And under such conditions, attention to construction of Roghun and other waterworks facilities should concentrate on long-term regulation rather than as a source of electric energy. Our calculations bring out clearly that under a long-term operation regime the Roghun reservoir does not operate practically at NMOL 1240, and it is important for the region to have a reservoir at NMOL 1290, which allows optimal regimes for all players of the Amudarya water-management system. Scenario "A" for the Amudarya basin means putting into operation Roghun HEPS at NMOL 1,290 m and an energy-oriented operation regime of Roghun and Nurek reservoirs, with the economic damage to basin's water sector at 174 M$/year. This is comparable to the cost of electricity generated by Roghun HEPS (see Tables 2 and 3 and Dukhovny and Sorokin 2007). When water use follows scenario "A" but a combined (energy-irrigation) regime is introduced, i.e. Roghun is operated in long-term energy-oriented regime and Nurek is operated in compensation irrigation-oriented regime, while long-term storage of Roghun is used to cover irrigation deficit, the average water deficit can be reduced to 1.5–2% (Table 4). The environmental damage would be about 10 M$/year, and over 50 years, there would be only one period (more than 2 years) of prolonged infringement of the Amudarya delta's interests.

Table 2 Production losses in irrigated agriculture and associated sectors in the Amudarya basin over 2005–2055 under scenario "A" and various options of flow regulation by reservoirs (M$/year)

Flow regulation options	Production losses due to natural and anthropogenic factors as compared to potential (planned) value
Without Roghun	95
Energy-oriented, NMOL 1290	174
Irrigation-oriented, NMOL 1290	38
Combined, NMOL 1290	76

Table 3 Impact of Roghun HEPS (normal maximum operating level 1290 m) on economic indicators of Amudarya basin countries development over 2005–2055 in scenario A (M$/year)

Scenarios of joint operation of Roghun HEPS and Nurek HEPS	Growth of production from irrigated agriculture and associated sectors	Cost of generated electric energy	Cumulative effect in the basin
Combined	19	195	214
Irrigation	57	188	245
Energy-generation	−79	195	116

Table 4 Indicators of water supply to irrigated agriculture in middle and lower reaches of the Amudarya under different operation options of Roghun (NMOL 1290 m) and Nurek reservoirs. Scenario "A," simulation period 2007–2055

	Reservoir operation options	Q-ty of irregular years, %	Mean water shortage over the period, %	Max annual depth of water shortage, %
1	Energy-oriented	36	6	34
2	Combined	18	2	19

The particularly abrupt reduction of damage or even its avoidance is possible under the scenario "B" which envisages long-term regulation in a combined regime.

In assessing possible future flow regulation scenarios for the Amudarya basin and their consideration in "rules," particular attention should be paid to analysis of future operation regimes of the Dasht-e-Dzhum hydrosystem, taking into account its impact on the Pyandj River's natural runoff regime, which presently corresponds fully to irrigated agriculture's demand hydrograph in middle and lower reaches of the Amudarya River. Given the planned construction period of 11 years, the Dasht-e-Djum waterworks (useful capacity 10.2 km^3) should be expected to be operational after 2025. In the future, Afghanistan may require that its water allocation share be increased for socio-economic development of its northern part. This would somewhat change runoff regimes in the Pyandj River and the Amudarya River. While developing "rules," one may consider an option envisaged in the "Scheme of irrigation development in Northern Afghanistan," with an additional supply of 3.6 km^3/year along the river.

As to siltation of reservoirs and probable losses of regulating capacities, the main siltation factor in the Amudarya basin for the Roghun and Nurek waterworks is the dynamics of future liquid and sediment runoffs in the Vaksh River, and is dependent on climate scenarios. For the in-stream reservoir of Tuyamuyun waterworks, along with inflow to the reservoir (dependent on climate scenario and basin development scenario), the operation regime of the reservoir, recommended by SANIIRI as flushing (scenario "B") or non-flushing (see Table 5), has considerable impact on siltation.

According to design developments, optimal irrigation and energy use in the Syrdarya basin can be achieved by putting into operation new HEPS upstream of the Toktogul waterworks (cascade of Kambarata HEPS), free of irrigation limitations and operating under a regime of seasonal energy compensators. Water-energy modeling demonstrates the efficiency of given measures for the basin as a whole, provided that HEPS cascades are operated not only to the benefit of energy users. Otherwise, damage for irrigated agriculture would only increase.

By assuming (scenario "B") that we would have agreed among the countries upon the methods for estimation of costs and benefits from the use of transboundary water resources, the objective of regional management (development) would be reduced to the search for general solutions that maximize national and regional interests (benefits). To this end, it is proposed, in high-water years, to allocate from the transboundary water resources a certain share for long-term flow regulation and use this share to cover water and energy shortages.

Table 5 Estimation of volumes of in-stream reservoirs in the Amudarya basin, scenario "B" (data from NTC "Toza Darye," Project "Jayhun" 516761 INCO)

Year	Nurek	Tuyamuyun	Roghun	Total
	Volume of siltation (km^3)			
1972	0	–	–	0
1978	0.7	0	–	0.7
2007	2.6	1.1	–	3.7
2025	3.2	1.5	0.8	5.5
2050	3.5	1.8	2.8	8.1
	Reservoir capacities, considering siltation (km^3)			
1972	10.5	–	–	10.5
1978	9.8	7.8	–	17.6
2008	7.9	6.7	–	14.6
2025	7.3	5.3	12.5	25.1
2050	7.0	6.0	10.5	23.5

Under scenario "B," the created ecosystems are to be sustained through water left after water limits allocation (given that minimum environmental inflows to Prearalie and the Aral Sea are kept by all means in any flow probability periods). This water is to be distributed between national water bodies (Arnasay, additional supply to Prearalie lakes and the Aral Sea) on a contractual basis. Operation regimes for interstate reservoirs are built in scenario "B" based on national requests (for accumulation in and releases from reservoirs), taking into account obligations regarding long-term regulation and current compensation of costs (damages) occurring when national regimes are changed to reach regional trade-offs.

Optimistic Scenario

Achievement of those objectives will allow water shortages in the region to be minimized and release water resources for the Aral Sea. Scenario "A" envisages the stabilization of productivity and demand for water from transboundary rivers and the development, mainly, through capacities of local water sources. According to this scenario, the water deficit for irrigated agriculture in the Syrdarya basin (average for 50 years) is estimated at 6% of limit, with a maximum of up to 20% in some years. For the Amudarya basin, this is 3.6%, with a maximum of 20% by states and 28% by provinces.

Environmental requirements for the Amudarya are met in high-water and normal years, while for the Syrdarya, in all options over a year as a whole. Scenario "B" stipulates that by 2050, 80% of potential land productivity would be achieved and off-takes from transboundary rivers would be reduced through water conservation. This would allow efficient allocation of free water in order to ensure long-term storage and water supply to Prearalie. According to this scenario, water deficit for irrigated agriculture in the Syrdarya and Amudarya basins will not be observed after

2030, and environmental requirements (sanitary releases along river channels, water supply to the system of lakes, river deltas and Arnasay) will be met.

The hydropower deficit for Kyrgyzstan would be compensated by Uzbekistan and Kazakhstan in some years and will not exceed 1.8 billion kwh (13% of power generation by HEPS in Kyrgyzstan).

Estimated inflow to Prearalie from Syrdarya and Amudarya Rivers as an average over 2007–2025 is shown in Fig. 4 for the scenario "B." The figure shows: average inflow by season, average river water salinity in scenario "B" and the changes in given values compared to those obtained for scenario "A," as well as options of water level stabilization in the Northern body of the Aral Sea and discharges into Large Aral. It is assumed that in scenario "B" because of implemented water-conservation policies, the following water-use efficiency indicators will be achieved: unit water use – 9,4 thousand m^3/ha in irrigation and 0,08 thousand m^3/person/year for population. The population growth rates will decrease to 0.98% by 2020. The irrigated area will stabilize at 8.5 million ha. The expansion of irrigated area is expected mainly after 2010. The total water consumption will be about 90 km^3/year, including 80 km^3/year for irrigation.

According to simulations, stabilization of the Northern Aral Sea at the level of 47m is possible under all scenarios of inflow to the Sea, the only difference is in the periods of stabilization: after 2040 in scenario "A"; and, after 2020 in scenario "B". The option of water supply to the Large Aral Sea under scenario "A" and the existing water infrastructure in South Prearalie lead to maintaining water level at 25 m in Eastern Bowl and lowering the level in Western Bowl down to 20 m by 2020.

The water supply option under scenario "B" leads to periodic junction and separation of the Large Sea's bodies, with the mean annual level of 28 m. Under the scenario "B" and the hypothetical option developed within the

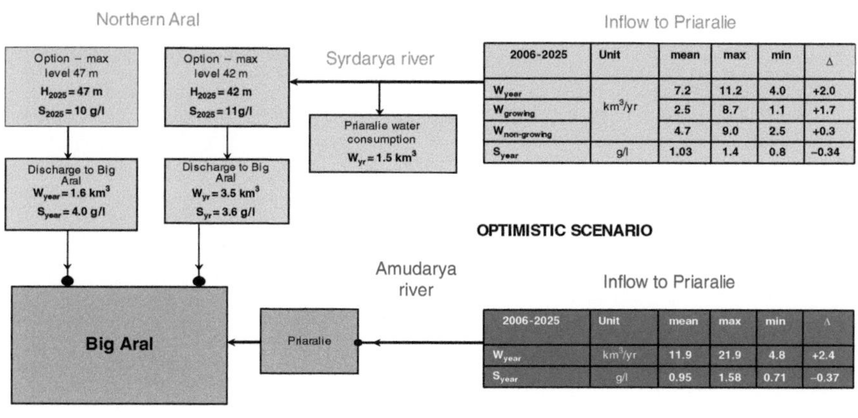

Fig. 4 Scheme of water inflow and use in Prearalie in the scenario "B" Water conservation and efficient water use are priority objectives at present and for the future

framework of the INTAS Project 0511, it is expected that supply to Western Bowl would be implemented through the system of Amudarya-Sudochie-Adjibay. The Eastern Bowl is fed only during overflows from the Western Bowl plus water releases from the Northern Aral. In Western Bowl, water level would be 29–31 m, with the short-term minimum of 28 m and the maximum of 32 m. Eastern Bowl would be stabilized at 26–27 m BS.

References

Dukhovny V.A., Sorokin A.G., Stulina G.V. Should we think about adaptation to climate change in Central Asia? Adaptation to climate change: regional challenges in light of world experience. SIC ICWC, Tashkent, 2008, p. 14–24.

Dukhovny V.A., Sorokin A.G. Assessment of Roghun reservoir impact on runoff regime in the Amudarya River. SIC ICWC, Tashkent, 2007, 128 p.

Food Security: Quantity and Quality Matters for Eating Good Bread in Uzbekistan

Kienzler Kirsten, Nazar Ibragimov, and John P.A. Lamers

Introduction

Food security is more than producing sufficient amounts of food. Food security must also cope with demands for the quality of the staples and its processed products. Although the latter aspect often is considered of less importance, the efforts to meet the quality demands are nevertheless much less than to meet the quantities needed. Proper cultivation practices by farmers can already support meeting quality standards as is illustrated in the case of winter wheat (*Triticum aestivum* L.) production in Khorezm, Uzbekistan.

Winter wheat is the most important cereal grown for making bread in Central Asian Uzbekistan. Whereas worldwide about 65% of the produced wheat originates from irrigated agriculture, in Uzbekistan virtually all winter wheat is cultivated on irrigated land (FAO and WFP 2000). Irrigated winter wheat production is a comparatively new practice in the country and became widespread in the mid-1990s as a result of the National Program on Self-Sufficiency in Foods.

During the time of the Soviet Union, Uzbek SSR imported wheat and flour because the domestic production was limited to about 200,000 ha located in rainfed areas with an average yield of about 0.5–1.0 t ha^{-1} (Kamilov et al. 2002). In 1996, it became a declared policy in Uzbekistan to decrease its dependency from wheat imports, and the domestic cultivation of this crop gained an enormous momentum. Consequently, during the past 13 years, the area under irrigated cereals increased around five times from 221,000 ha to about 1 million ha, and gross yields rose from 1.54 in 1991 to 4.48 t ha^{-1} in 2008 (FAOSTAT 2009).

Gluten and protein content are key, albeit not the only, indicators for baking quality and food quality of kernels and therefore are used for assessing the baking quality of

K. Kirsten (✉)
International Center for Agricultural Research in the Dry Areas (ICARDA), Regional Office for Central Asia and the Caucasus (CAC), P. O. Box 4564, 100000, Tashkent, Uzbekistan
e-mail: k.kienzler@cgiar.org

C. Madramootoo and V.A. Dukhovny (eds.), *Water and Food Security in Central Asia*, NATO Science for Peace and Security Series C: Environmental Security,
DOI 10.1007/978-90-481-9974-7_13,

wheat flour (Gupta et al. 1992, Shewry et al. 1995). Common quality criteria for wheat include high flour protein, high water absorption, good dough extensibility and tolerance to mixing, and high loaf volume (Schofield 1994, Shewry et al. 1995, Bruckner et al. 2001). While providing viscosibility and elasticity, the gluten is responsible for the functionality of wheat flour and eases the processing into different food stuffs (Shewry et al. 2002). A balance between viscosibility and elasticity is important, as highly extensible or too strong gluten will not produce the desired voluminous, fluffy, evenly pored dough (Shewry et al. 1995). Also, a low protein content will cause problematic starchy kernels (Fowler et al. 1989). Wheat kernels at maturity, for example, may contain between 8–20% protein (Johnson et al. 1973, Farrer et al. 2006). The optimal level for bread making is provided at protein contents of 11.5–14.0% (Oliver 1988, Panozzo and Eagles 2000).

However, a recent survey confirmed that the milling and baking industry is well aware of the low gluten content and weak gluten character of the presently grown winter wheat varieties in Uzbekistan, and acknowledges that the present concentrations are unsuitable to produce high quality Tandyr bread[1] (Abugalieva et al. 2003a). Furthermore, the domestically produced winter wheat has a gluten content of less than 20%, which is an overall low gluten quality, and often has starchy kernels (Abugalieva et al. 2003a) caused by low protein content (Fowler et al. 1989).

Hard wheat varieties grown in semi-arid environments tend to have higher protein contents than soft wheat varieties (Habernicht et al. 2002). Roughly, for bread wheat, the German classification for instance distinguishes three protein and four gluten classes (Raiffeisen 2008): low protein (10.5%), medium protein (12.5%) and high protein (16.5%) content; and little gluten (<20%), low gluten (20–23%), medium gluten (24–27%) and high gluten (>28%) content. The official Soviet standard for hard wheat provides a more detailed system for protein and gluten (Table 1) (Abugalieva et al. 2003b).

The content of protein and gluten in wheat kernels can be managed by targeted nitrogen (N) fertilizer applications during wheat cultivation (e.g., Farrer et al. 2006). Also, kernel weight can be improved for instance by applying higher N-fertilizer rates (e.g., Alaru et al. 2003). On the other hand, this should not be understood as a

Table 1 Wheat protein and gluten classification for hard and medium winter wheat varieties as determined during the Soviet Union Era (Abugalieva et al. 2003b)

	Strong wheat types			Value wheat types	Wheat filler		Weak wheat
	Excellent improver	Good improver	Satisfactory improver		Good	Satisfactory	
Protein content,% (not less than)	16.0	15.0	14.0	13.0	12.0	11.0	8.0
Kernel gluten content, % (not less than)	32.0	30.0	28.0	25.0	24.0	22.0	15.0

[1] Tandyr bread is locally made traditional flat bread from wheat flour.

blind recommendation since there is a natural limit above which increasing protein content is associated with decreasing yields, as protein content and yield are negatively correlated (e.g., Olson et al. 1976, Alaru et al. 2003). Once the N-response curves for both protein content and yield for the different varieties are known, breeders may narrow the ratio of yield-to-protein content beyond the potential of each variety by targeted selections of particular improved varieties (e.g., Terman 1979, Lanning et al. 1994, Ortiz-Monasterio et al. 1997, Fowler 2003).

In addition to increasing N rates, the appropriate timing of N application significantly affects protein content (e.g., Farrer et al. 2006, IFA 2006). For instance, when the N application until the reproduction phase is delayed (i.e. the periods of maximum N assimilation or of maximum carbohydrate formation) e.g., prior to anthesis/flowering (Zadoks-60, Feekes-10.51 (Zadoks et al. 1974)), or after anthesis/at milk formation stages (Zadoks-73, Feekes-11.1 (Zadoks et al. 1974)), increased protein contents of the kernels can be achieved without affecting the vegetative growth. In fact, the efficiency of acquired N after anthesis was higher than for earlier N applications, as the partitioning of the absorbed N to the kernels was more effective (Wuest and Cassman 1992). On the other hand, during this growth stage the mineral N content in the soil is usually low (Smith and Whitfield 1990), which also may impact N efficiency.

Given the present production practices in Uzbekistan, there is room for increasing the quality of wheat kernels and flour. Revising and adjusting N-fertilizer management strategies is a first step towards improved quality, once quality becomes a priority. This study reports on findings from various field experiments during 2003–2006 focusing on the effect of N fertilization on the baking quality of winter wheat grain in the Khorezm region in Northwest Uzbekistan.

Materials and Methods

Field experiments with the winter wheat variety Kupava R2 were conducted in the Khorezm region from 2003–2006 within the framework of the German-Uzbek ZEF/UNESCO project. This region has a typical continental semi-arid climate. Mean annual temperature is 13.7°C with a minimum in February (−9°C) and a maximum in June/July (40°C). Mean annual rainfall does not exceed 100 mm (75 mm in 2004, Khiva Meteorological Station) with maximum precipitation in April and November (Fig. 1). Evaporation of about 1,400–1,600 mm exceeds precipitation throughout the year.

The Khorezmian soils are of alluvial origin. According to Russian and Uzbek literature, the main soil type found in the region is the so-called irrigated alluvial meadow soil covering about 60% of the area (Rasulov 1989, Djumaniyazov 2006).

The experimental layout was a complete randomized block design with three or four replications. Nitrogen was applied in three split applications at total rates of 0, 20, 80, 120, 160, 180, 240 and 300 kg N ha^{-1}. Before seeding, 20% of the total N rate was applied. Both at booting and tillering, 40% of the total rate was applied, each time directly followed by irrigation ("3 split applications"). One set of treatments ("4 split applications") of the fertilizer rate 120 kg N ha^{-1} received an

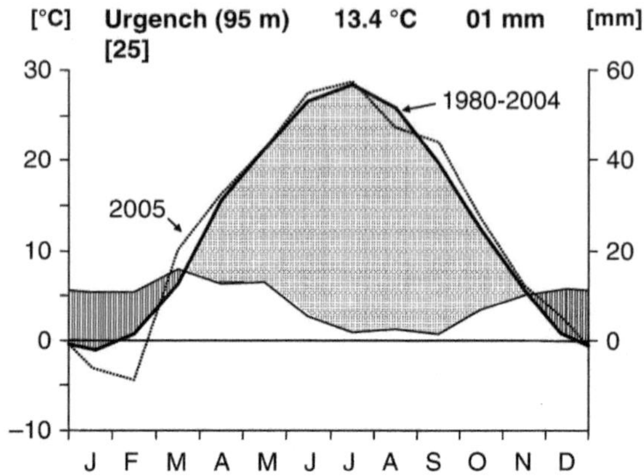

Fig. 1 Climate diagram for Urgench, Khorezm, Uzbekistan, according to Walter and Leith (1967)

additional N fertilizer application just before heading, i.e. 30% of the N rate at tillering and booting and 20% at heading.

The seeding rate was 250 kg ha^{-1}. Three sub-plots each of 1 m^2 were harvested to determine key yield components such as average kernel weight per m^2, spikes per m^2, weight of kernels per spike and 1,000-kernel weight (TKW). Additionally, fresh and dry (105°C) weight of total biomass, stems, spikes, and chaff, the length of plant and spikes, and plant density (number of plants with spikes per m^2) were measured for each m^2. The harvest index (HI) was calculated as the ratio of kernel weight to total biomass. Grain sub-samples were analyzed for gluten and protein content, transparency as well as gluten quality. The N concentration was determined using the Kjeldahl method. The crude protein concentration was computed based on the assumption that the N content of protein is 16%, or N content × 6.25.

The mean effect of treatment variables and their interactions were compared at $p < 0.1$ level of significance with the analysis of variance (ANOVA) procedure. LSD post hoc test compared individual treatment means.

Results

The findings confirmed the recommendations of the Uzbek Cereal Research Institute for Irrigated Land (MAWR 2000): a N-fertilizer application of between 160 and 180 kg N ha^{-1} gave highest wheat yield (Fig. 2).

According to national wheat quality classifications for bread-making, medium quality wheat should have a gluten content of more than 23% and a crude protein content of more than 13%. The regional wheat in 2003–2006 was classified as class 3, which is equivalent to soft wheat of a gluten quality value of 105–120. All wheat produced during the experiments, except four cases, was classified in the better classes 1 and 2. In comparison with the German protein classification for baking quality (Raiffeisen 2008), the protein class "low" (10.5%) was met when 120 kg N ha^{-1} was applied (Fig. 3). At the N rates of more than 180 kg N ha^{-1}, the kernels reached the

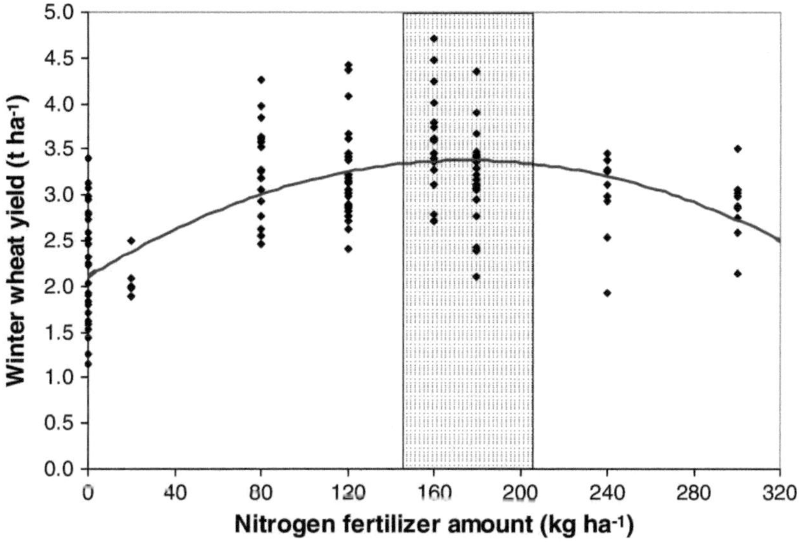

Fig. 2 Total Khorezmian winter wheat yields (t ha^{-1}) for the respective N rates (kg ha^{-1}) from 2003–2006. The regression curve for average yields for the respective N rate is made for visualization only. The *shaded areas* represent the official N-fertilizer recommendations for winter wheat (MAWR 2000)

protein class "medium" (12.5%). All wheat kernels receiving less than 120 kg ha^{-1} of N fertilizer were below the minimum requirements.

The winter wheat crude protein content was significantly higher in 2004 (13.1 ± 1.5%) and 2005 (12.4 ± 1.7%) compared to 2006 (10.2 ±1.3%) (Fig. 4). Also, the N rates made a difference at $p < 0.1$ (Table 2): Kernels with N rates of 240 and 300 kg ha^{-1} had significantly higher protein contents (14.1 ± 1.4 and 15.2 ± 1.8%) than those from the lower N rates.

Kernel gluten content was on average 22.1 ± 4.3%. Gluten content differed among years and the interaction year × N rate was significant. The N rate alone was not significant. The gluten content was significantly lower in 2004 (20.7 ± 5.7%) than in 2005 (23.7 ± 3.0%).

The kernels of all N rates had gluten concentrations higher than the threshold value (Raiffeisen 2008) of 20.0% gluten (Fig. 3, Table 2). The "medium" class (23.5%) was reached for wheat fertilized with 160–180 kg N ha^{-1} whereas higher N application rates did not bring the kernels to the highest gluten class.

The timing of the fertilizer application significantly influenced the crude protein content of the kernels. Wheat from the treatment with four split applications, which had received an additional N-fertilizer application at heading, had significantly higher crude protein contents (10.3 ± 0.8%) than that from those treatments that received the recommended (MAWR 2000) three split applications (Fig. 5).

The 1,000-kernel weight (TKW) of the Khorezmian wheat (Fig. 6) was generally higher than the 5-year average of 33 g in US wheat classified as soft red wheat (Gwirtz et al. 2007). This weight, however, differed according to the harvest year:

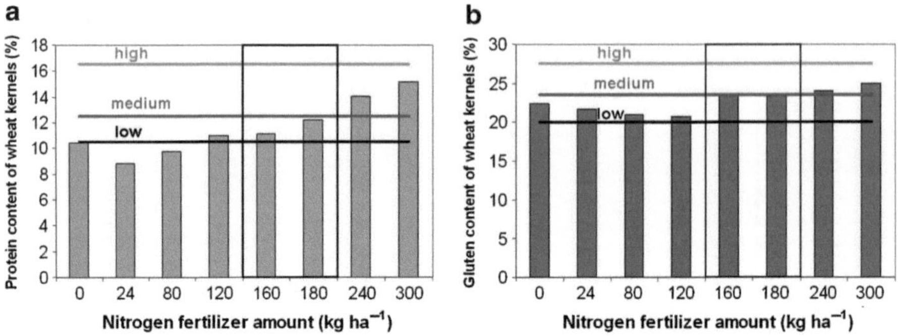

Fig. 3 Average crude protein and gluten content of Khorezmian winter wheat kernels (%) for the respective N rates (kg ha^{-1}) for 2003–2006. The horizontal bars indicate the three protein and four gluten classes of the German classification (Raiffeisen 2008). The *shaded areas* represent the official N-fertilizer recommendations for winter wheat (MAWR 2000)

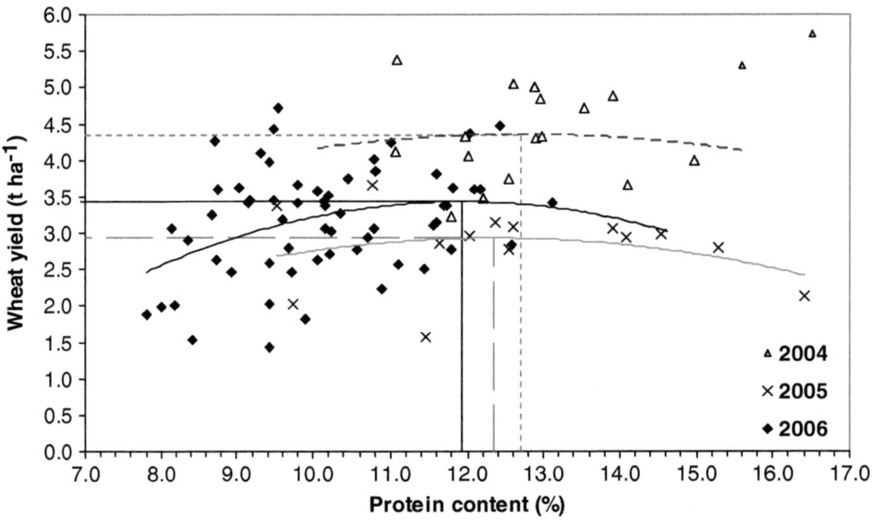

Fig. 4 Relationship of mean crude protein content (%) and mean kernel yield (kg ha^{-1}) of Khorezmian winter wheat grown with different levels of N in 2004–2006. The regression curves for the respective years are made for visualization only

kernels were significantly lighter in 2005 (34.1 ± 2.5 g) than in 2004 and 2006 (37.4 ± 2.1 g).

The TKW followed the trend of the yield; the response to the N rate can be described by a quadratic function. The TKW was significantly higher for N-160 (38.9 ± 1.6 g) than for the other N rates (Table 2). The lowest TKW was found for N-0 (33.3 ± 2.6 g). N fertilization management also had a significant effect on TKW since kernels of treatment receiving 4 split applications were 2 g heavier (39.3 ± 1.9 g) than those treatments receiving only 3 split applications (on average 37.5 g) (data not shown).

Table 2 Average winter wheat yield (t ha^{-1}), crude protein and gluten content (%) and 1,000-kernel weight (TKW, g) for 2004–2006. SE denotes standard error of the mean

	Wheat yield			Protein content			Gluten content			TKW		
N rate	Mean	SE	$p<0.1$	Mean	SE	$p<0.1$	Mean	SE	$p<0.1$	Mean	SE	$p<0.1$
kg ha^{-1}	t ha^{-1}			%	%		%	%		g	g	
0	**2.31**	0.18	a	**10.4**	0.5	Abc	**22.4**	1.3	a	**33.3**	0.5	A
24	**2.09**	0.14	a	**8.9**	0.9	A	**21.7**	2.5	a	**35.3**	0.6	Bc
80	**3.27**	0.13	bc	**9.8**	0.3	Ab	**21.0**	0.8	a	**37.2**	0.3	cd
120	**3.57**	0.14	c	**11.0**	0.2	Bc	**20.7**	1.0	a	**36.7**	0.4	bc
160	**3.60**	0.14	c	**11.1**	0.3	Bc	**23.5**	1.1	a	**39.0**	0.4	D
180	**3.93**	0.27	c	**12.3**	0.3	C	**23.0**	1.3	a	**35.5**	0.4	Bc
240	**3.98**	0.34	c	**14.1**	0.5	D	**24.0**	1.3	a	**36.3**	0.9	Bc
300	**2.60**	0.46	ab	**15.2**	1.3	D	**25.0**	5.0	a	**34.9**	1.0	Ab

Means with the same letter in the column are not significantly different according to the Tukey test ($P=0.1$)

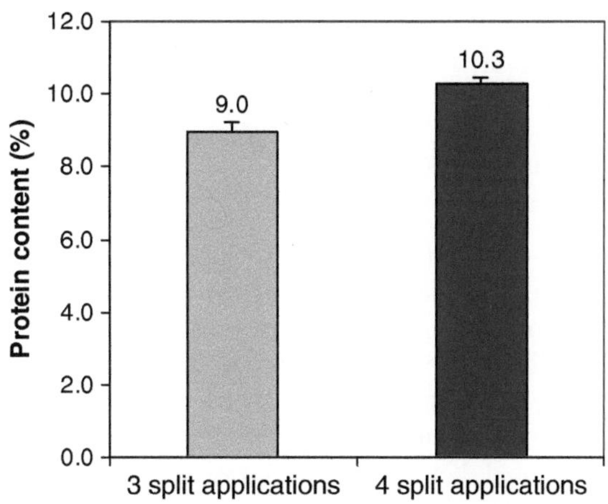

Fig. 5 Average crude protein content (%) of Khorezmian winter wheat kernels fertilized at a rate of 120 kg N ha^{-1}. Fertilizer rates were applied in three split applications as recommended (MAWR 2000) and with four split applications (one additional at heading). *Error bars* represent 1 standard error (SE)

Discussion

Grain yield and protein content in the wheat kernels increased with higher N rates. However, while the increase in kernel protein content was linearly related to N application rates, the relationship between yield and N application followed a quadratic function. Furthermore, although the curves of the regression were rather flat, a quadratic relationship between yield and protein content could be discerned. This indicates that the maximum yield of the variety Kupava R2 did not correspond to the highest protein content: at the highest N amendments tested maximum yields decreased again whereas protein content increased. The highest protein level of 15.2% was achieved by applying 300 kg N ha^{-1}, but at this N rate yields decreased

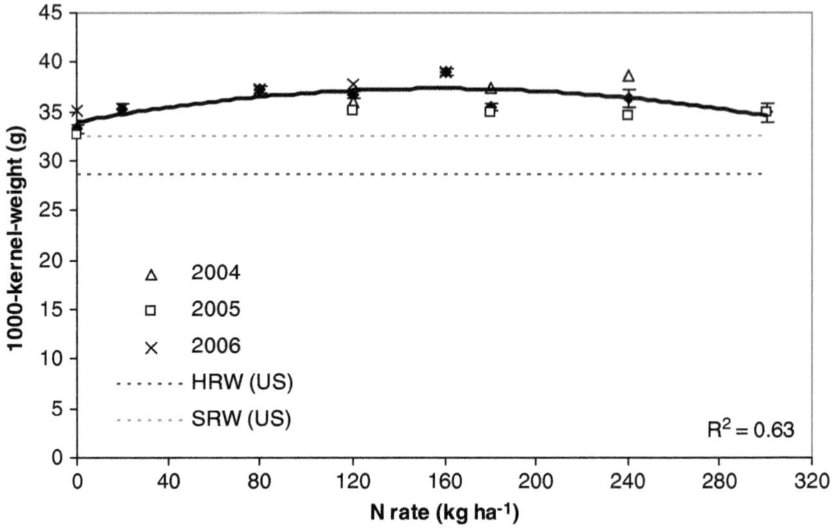

Fig. 6 Average 1,000-kernel weight of Khorezmian winter wheat (g) for the respective N rates (kg N ha^{-1}) for 2004–2006. *Error bars* represent 1 SE of the mean. The regression curve for average yields for the respective N rate is presented for visualization only. The *horizontal dotted lines* indicate the 5-year average of US wheat for hard red wheat (HRW) and soft red wheat (SRW) (Gwirtz et al. 2007)

from the maximum of 3.6 to approximately 2.7 t ha^{-1}. The protein content at the highest yield level (N rate of 181 kg N ha^{-1}) was only around 12.2%, just reaching the medium quality protein level (Raiffeisen 2008). A similar relationship was previously reported (Johnson et al. 1973), where the relationship between protein and yield increase for the wheat variety Lancer was quadratic as opposed to the variety Comanche that showed a linear relationship. Selles and Zentner (2001) attributed their monitored negative correlation between kernel protein and yield to water stress rather than N availability for the crop. This explains the recurrently mentioned need to breed and select for wheat varieties with high quality and yield potential so as to increase the guarantee of high yields at acceptable quality levels (e.g., Johnson et al. 1973, Cox et al. 1985, Calderini et al. 1995, Ortiz-Monasterio et al. 1997, Fowler 2003).

Significant year and location effects on yield, protein and gluten levels have been observed previously (e.g., Fowler et al. 1989, Lloveras et al. 2001, Alaru et al. 2003, Farrer et al. 2006). These effects were considered of equal, if not of more importance than the influence of the wheat genotypes (e.g., Fowler 2003). Especially planting dates, seasonal temperatures, irrigation timing and related water stress during spring up to anthesis are known to influence the tiller density, N accumulation, seed size and wheat quality (Fowler et al. 1990, Farrer et al. 2006).

The monitored (slight) response of TKW to N rates agrees with findings by Alaru et al. (2003) but contrasts to the results of Frederick et al. (2001) who noted that the TKW was only marginally correlated with overall yield. They attributed this to the breeder's selection for higher kernel number per square meter rather than for heavier kernels. Also, Eck (1988) and Badaruddin et al. (1999) found insignificant differences in seed weight among N treatments. The TKW was affected more by warm, dry weather conditions and water stress during grain filling (e.g., Frederick et al. 2001). Water-stress induced premature ripening resulted in lower seed numbers and kernel weights, and thus reduced yields (e.g., Eck 1988).

Overall, officially recommended N-fertilizer rates of 150–180 kg N ha^{-1} (MAWR 2000) can be regarded as acceptable for wheat production. However, the findings of the protein and gluten concentrations show that the Khorezmian winter wheat variety Kupava R2 can meet the criteria only of a satisfactory to good wheat filler and flour thickener of low to medium quality (Oliver 1988, Abugalieva et al. 2003b, Raiffeisen 2008). There is thus much room for improvement, in particular by increasing N use efficiencies through judicious N application strategies. The present results showed that wheat receiving an additional N rate at heading yielded highest protein content in the kernels. Late N applications thus could be one option to increase N concentrations in the wheat kernels and in turn improve quality. A supplemental fertilization at the Zadoks-60/Feekes-10.51 or at the Zadoks-73/Feekes-11.1 stages is a common management strategy to raise protein levels in specific environments (e.g., Ottman et al. 2000, Woolfolk et al. 2002, IFA 2006).

On the other hand, an increase in quality due to extra N applications obviously comes at the expense of higher costs to the producers and therefore would need to be compensated for by the market, buyers or consumers. In the absence of regulating mechanisms, it is unrealistic to expect that producers maximize protein production, when they can keep their focus on maximizing yields at lower costs. However, given the recently experienced soaring prices for wheat and the present necessity for Uzbekistan to import better quality flour to mix with the domestically produced wheat to upgrade the baking quality (Rudenko 2008), any progress in increasing kernel quality could become on opportunity for the future. Higher wheat quality could not only contribute to the declared direction of wheat self-sufficiency, but also could open room for exporting high quality wheat flour once domestic demands are satisfied.

Conclusions

The present recommendations for N-fertilizer applied to wheat are conducive to achieve high yields and meet the domestic wheat demands and in turn contribute to food security. A next challenge is the improvement of the present low baking quality. The presented quality data classify wheat variety Kupava as a satisfactory to good wheat filler but low to medium quality flour thickener. Late N applications at for instance anthesis/heading significantly increased the fertilizer-N uptake efficiency and crude protein content in the kernels while gluten content was less affected by higher N rates. Furthermore, crude protein content and yield were negatively related for the variety examined, i.e., the maximum protein content did not coincide with maximum yield, showing the need for breeding wheat varieties with higher quality and yield potential suitable for the irrigated lowland areas of the region. Yet, at present farmers are more encouraged to produce higher yields at lower (fertilizer) costs rather than producing concurrently higher protein and quality grain. To improve the quality of Khorezmian wheat, and, subsequently, also increase its market and export value, farmers could be encouraged by offering adequate reimbursements for high quality rather than for quantity only. Also, a regular monitoring of the crude protein content as well as of other quality aspects of wheat ought to be part of the fertilizer recommendations. This will make it possible to optimize N and water management.

Acknowledgements This study was funded by the German Ministry for Education and Research (number 0339970A-E), by the Ministry for Schools, Science and Research of the State of Northrhine-Westfalia, Germany, and the Robert Bosch Foundation. We thank the participating Khorezmian farmers for their support.

References

Abugalieva A., Atli A., Murgounov A., Massalimov A., Meng E. C. H. and Peña R. J. 2003a. Milling, baking and grain factors influencing quality of tandyr bread (nan) in Central Asia. In: GTZ-CIMMYT (ed.), Quality of wheat kernels in Central Asia. GTZ-CIMMYT, Almaty. p.5–22.

Abugalieva A. I., Bedoshivili D. and Morgunov A. I. 2003b. Main characteristics of baking quality of Kazakh winter wheat kernels in comparison with requirements of the world market. Description of the existing system of wheat quality classification and its comparison with the international systems (US, Europe, Canada, Australia). In: GTZ-CIMMYT (ed.), Quality of wheat kernels in Central Asia. GTZ-CIMMYT, Almaty. p.23–30.

Alaru M., Laur Ü. and Jaama E. 2003. Influence of nitrogen and weather conditions on the grain quality of winter triticale. Agron Res 1: 3–10.

Badaruddin M., Reynolds M. P. and Ageeb O. A. A. 1999. Wheat management in warm environments: Effect of organic and inorganic fertilizers, irrigation frequency, and mulching. Agron J. 91(6): 975–983.

Bruckner P. L., Habernicht D., Carlson G. R., Wichmanc D. M. and Talbert L. E. 2001. Comparative bread quality of white flour and whole grain flour for hard red spring and winter wheat. Crop Sci 41: 1917–1920.

Calderini D. F., Torres-Leon S. and Slafer G. A. 1995. Consequences of wheat breeding on nitrogen and phosphorus yield, grain nitrogen and phosphorus concentration and associated traits. Ann Bot 76(3): 315–322.

Cox M. C., Qualset C. O. and Rains D. W. 1985. Genetic variation for nitrogen assimilation and translocation in wheat. II. Nitrogen assimilation in relation to grain yield and protein. Crop Sci 25(3): 435–440.

Djumaniyazov A. 2006. Literature review of soils in Khorezm. Internal report. ZEF, Urgench, Uzbekistan, 34 p.

Eck H. V. 1988. Winter wheat response to nitrogen and irrigation. Agron J 80(6): 902–908.

FAO and WFP 2000. Special report: FAO/WFP Crop and food supply assessment mission to the Karakalpakstan and Khorezm regions of Uzbekistan. FAO/WFP, Rome, 17 p.

FAOSTAT 2009. Agriculture data, Agricultural production. http://www.fao.org, last accessed 22.08.2009

Farrer D. C., Weisz R., Heiniger R., Murphy J. P. and White J. G. 2006. Minimizing protein variability in soft red winter wheat: Impact of nitrogen application timing and rate. Agron J 98(4): 1137–1145.

Fowler D. B. 2003. Crop nitrogen demand and grain protein concentration of spring and winter wheat. Agron J 95(2): 260–265.

Fowler D. B., Brydon J. and Baker R. J. 1989. Nitrogen fertilization of no-till winter wheat and rye. II. Influence on grain protein. Agron J 81(1): 72–77.

Fowler D. B., Brydon J., Darroch B. A., Entz M. H. and Johnston A. M. 1990. Environment and genotype influence on grain protein concentration of wheat and rye. Agron J 82(4): 655–664.

Frederick J. R., Bauer P. J. and Busscher W. J. 2001. Grain yield and yield components of doublecropped winter wheat as affected by wheat and previous soybean production practices. Crop Sci 41: 778–784.

Gupta R. B., Batey I. L. and MacRitchie F. 1992. Relationships between protein composition and functional properties of wheat flours. Cereal Chemistry 69: 125–131.

Gwirtz J. A., Willyard M. R. and McFall K. L. 2007. Wheat quality in the United States of America. In: Popper L., Schäfer W. and Freund W. (eds.), Future of Flour – A Compendium of Flour Improvement. Agrimedia, Ahrensburg, pp.17–42.

Habernicht D. K., Berg J. E., Carlson G. R., Wichman D. M., Kushnak G. D., Kephart K. D. et al. 2002. Pan bread and raw Chinese noodle qualities in hard winter wheat genotypes grown in water-limited environments. Crop Sci 42: 1396–1403.

IFA 2006. World fertilizer use manual. Paris, International Fertilizer Industry Association (IFA), 632 p.

Johnson V. A., Dreier A. F. and Grabouski P. H. 1973. Yield and protein responses to nitrogen fertilizer of two winter wheat varieties differing in inherent protein content of their grain. Agron J 65(2): 259–263.

Kamilov B., Ibragimov N., Evett S. and Heng L. 2002. Use of neutron probe for investigations of winter wheat irrigation scheduling in automorphic and semi-hydromorphic soils of Uzbekistan. Conference Proceedings "International Workshop on Conservation Agriculture for Sustainable Wheat Production in Rotation with Cotton in Limited Water Resource Areas", Tashkent, Uzbekistan, October 13–18, 2002.

Lanning S. P., Talbert L. E., McGuire C. F., Bowman H. F., Carlson G. R., Jackson G. D. et al. 1994. Registration of 'McNeal' wheat. Crop Sci 34: 1126–1127.

Lloveras J., Lopez A., Ferran J., Espachs S. and Solsona J. 2001. Bread-making wheat and soil nitrate as affected by nitrogen fertilization in irrigated mediterranean conditions. Agron J 93(6): 1183–1190.

MAWR 2000. Agrotechnology of cereal production in irrigated and non-irrigated areas (in Russian). HAYOT, Andijan, Uzbekistan, 46 p.

Oliver J. R. 1988. Why is protein so important? Conference Proceedings "The Impact of Change – The Need to Adapt. Riverina Outlook Conferences", Australia, The Regional Institute, www.regional.org.au/au/roc/1988

Olson R. A., Frank K. D., Deibert E. J., Dreier A. F., Sander D. H. and Johnson V. A. 1976. Impact of residual mineral N in soil on grain protein yields of winter wheat and corn. Agron J 68(5): 769–772.

Ortiz-Monasterio R. J. I., Penna R. J., Sayre K. D. and Rajaram S. 1997. CIMMYT's Genetic Progress in Wheat Grain Quality under Four Nitrogen Rates. Crop Sci 37 (3): 892–898.

Ottman M. J., Doerge T. A. and Martin E. C. 2000. Durum grain quality as affected by nitrogen fertilization near anthesis and irrigation during grain fill. Agron J 92(5): 1035–1041.

Panozzo J. F. and Eagles H. A. 2000. Cultivar and environmental effects on quality characters in wheat. II. Protein. Australian Journal of Agricultural Research 51(5): 629–636.

Raiffeisen 2008. Ackermanager - Weizen. Qualitätskriterien von Weizen. http://www.raiffeisen.com/pflanzen/ackermanager/weizen_html, last accessed 07.07.2008

Rasulov A. M. 1989. Soils of the USSR cotton belt (in Uzbek). In: Cotton Compendium. Uzbek Soil Science Institute, Tashkent, pp. 26–36.

Rudenko I. 2008. Value chains for rural and regional development: The case of cotton, wheat, fruit and vegetable value chains in the lower reaches of the Amu Darya River, Uzbekistan.

Wirtschaftswissenschaften (Economy), Hannover, ZEF Bonn/Gottfried Wilhelm Leibniz Universität Hannover, 235 p.

Schofield J. D. 1994. Wheat proteins: structure and functionality in milling and breadmaking. In: Bushuk W. and Rasper V. F. (eds.), Wheat production, properties and quality. Chapman & Hall, London, p.72–106

Selles F. and Zentner R. P. 2001. Grain protein as a post-harvest index of N sufficiency for hard red spring wheat in the semiarid prairies. Can. J. Plant Sci 81(4): 631–636.

Shewry P. R., Tatham A. S., Barro F., Barcelo P. and Lazzeri P. 1995. Biotechnology of breadmaking: Unraveling and manipulating the multi-protein gluten complex. Bio/Technology. 13: 1185–1190.

Shewry P. R., Halford N. G., Belton P. S. and Tatham A. S. 2002. The structure and properties of gluten: an elastic protein from wheat grain. Philos Trans Roy Soc B Biol Sci 357 (1418): 133–142.

Smith C. J. and Whitfield D. M. 1990. Nitrogen accumulation and redistribution of late applications of 15N-labelled fertilizer by wheat. Field Crop Res 24(3–4): 211–226.

Terman G. L. 1979. Yields and protein content of wheat grain as affected by cultivar, N, and environmental growth factors. Agron J 71(3): 437–440.

Walter H. and Leith H. 1967. Klimadiagramm-Weltatlas. 3rd edition. Jena, VEB Gustav-Fischer-Verlag p.

Woolfolk C. W., Raun W. R., Johnson G. V., Thomason W. E., Mullen R. W., Wynn K. J. et al. 2002. Influence of late-season foliar nitrogen applications on yield and grain nitrogen in winter wheat. Agron J 94(3): 429–434.

Wuest S. B. and Cassman K. G. 1992. Fertilizer-nitrogen use efficiency of irrigated wheat: II. Partitioning efficiency of preplant versus late-season application. Agron J 84 (4): 689–694.

Zadoks J. C., Chang T. T. and Konzak C. F. 1974. A decimal code for the growth stages of cereals. Weed Research. 14: 415–421.

Water Resources: The Basis for the Socio-Economic Development in the Lower Reaches of the Amudarya: A Case Study of Karakalpakstan

Kurbanbaev Erejep, Karimova Olga Yushevna, and Kurbanbaev Sagit

The fate of the Priaralie region and further agricultural development in the lower reaches of Amudarya depend entirely on the water policies of Tajikistan, Uzbekistan and Turkmenistan. The emergence of a critical water situation during low water years (2000–2003), when the water availability in the upstream and middle-stream of the river fluctuated in the range of 80–85%, and declined to 16% in the northern regions of the Republic of Karakalpakstan, is evidence of this dependence. In low water years, ecological objects and agriculture in the lower reaches of Amudarya suffered from the lack of water in low-water years, whereas in the high water periods a critical situation emerged due to dam and structures breakages in the river delta (Fig. 1).

The main causes of the emergence of such situations are:

- Lack of integrated policy on water resources management along the Amudarya River
- Inconsistent operation regime of large reservoirs, such as Nurek and Tuyamuyun
- Lack of accurate accountability of water withdrawal volumes along the river
- No recognition of ecological water requirements of the river delta and the Aral Sea by the states

The reduction in inflow of the river run-off to the Priaralie region has caused huge socio-economic damage, associated with deterioration of the environmental situation and loss from agricultural production resulting from the shrinkage of crop areas and decline in crop yield.

K. Erejep (✉)
Karakalpak Branch, Central Asian Research Institute of Irrigation (SANIIRI),
12 Dustlik Str, Nukus 742015, Republic of Uzbekistan
e-mail: info@icwc-aral.uz

C. Madramootoo and V.A. Dukhovny (eds.), *Water and Food Security in Central Asia*, NATO Science for Peace and Security Series C: Environmental Security,
DOI 10.1007/978-90-481-9974-7_14,

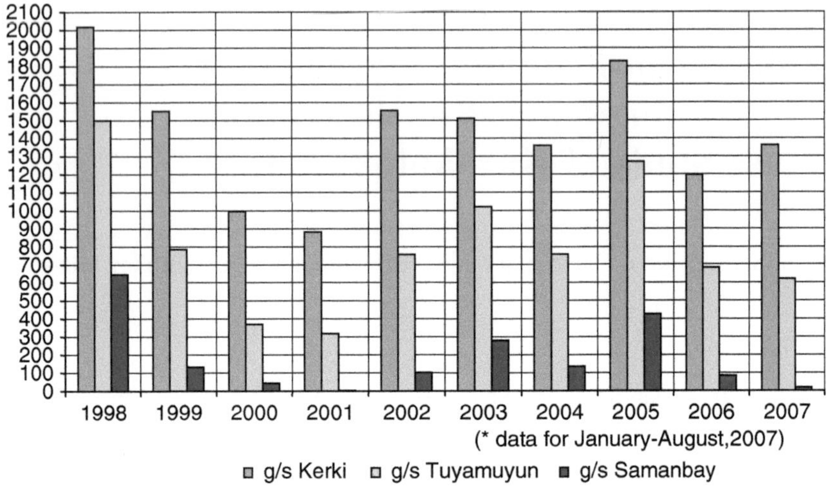

Fig. 1 Discharge fluctuations in the Amudarya River, measured at certain gauging stations for the period 1998–2007*

Socio-Economic Damage Caused by Aggravation of the Ecological Situation

The reduction in water supply for the lower reaches has resulted in the following unfavorable outcomes:

- Loss of the Aral Sea
- Losses in fisheries, muskrat breeding and livestock sector
- Irreversible character of natural changes due to the desertification of the river delta
- Creation of a new desert area on the dried bottom of the Aral Sea, which results in salt and dust transport to irrigated lands
- Loss in recreational value of the sea

Decrease of the Aral Sea Level

As of June 2007, the Aral Sea level has dropped by 24 m (Fig. 2). No water was discharged to the sea, even in average water availability years (2006–2007).

Although great efforts and policies on the stabilization of the sea level have been undertaken, decisions concerning the guaranteed water supply to the sea remain difficult to fulfill. In these circumstances, it should be acknowledged that if water allocation and use policies in the Aral Sea basin remain the same, the sea level will continue to fall. At present, the water is discharged into the sea on the

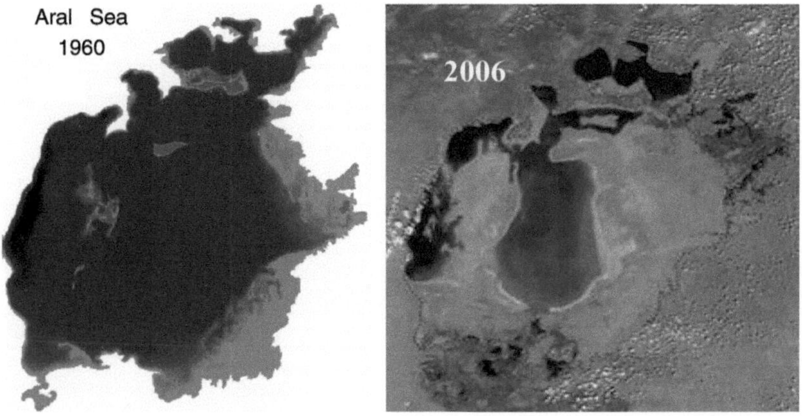

Fig. 2 The Aral sea basin in 1960 and 2006

residual principle. This means that one can expect an entire extinction of the sea in the future if no measures are undertaken (The Aral Sea Expedition - 2004, EC IFAS, Dushanbe, 2005).

A two-stage approach should be applied in order to deal with the Aral Sea problem: In the initial stage (stage I), a project should be developed with the involvement of interested donors (IFAS, GEF and others.). Meanwhile, feasibility studies aiming at determining the future of the Aral Sea are to be carried out based on project outcomes. If the project justifies saving the sea in a reduced, yet stable basin, then it can proceed to Stage II which aims at reaching an interstate consensus on the shared contribution of each republic to the amount of water that will be supplied to the sea. Since the future of the Aral Sea entirely depends on the policies of Central Asian countries, it will be necessary to clarify the following positions:

- Whether the Aral Sea is considered a problem by all Central Asian states, and if it should be solved
- Will all Central Asian countries agree to direct water saved in their territories to the Aral Sea
- Will Central Asian countries agree to allow BWO "Amudarya" and BWO "Syrdarya" to take full responsibility for guaranteed water supply (according to limits) to Priaralie and the sea

Losses in Fisheries and Muskrat Breeding

Before 1970, there were 34 species of fish inhabiting in the Aral Sea. Of these, more than 20 species were caught. At the same time, the sea supplied more than 4500 ton of commercial fish, among which valuable species were prevailing. They included thorn, pikeperch, silurus, aral barbel, and carp. The Aral Sea ranked first in the former Soviet Union in terms of fish catch. Currently it has lost its importance as a

Table 1 Fish catch in the Republic of Karakalpakstan from 1960 to 2006

Year	Weight ton	Year	Weight ton
1985	2460	2002	200
1990	2090	2003	130
1995	2090	2004	330
2000	1100	2005	440
2001	550	2006	610

Table 2 Muskrat skin production in the Republic of Karakalpakstan

Year	ths. units	Year	ths. units
1957	1,130	2003	0
1999	1.9	2004	0
2000	2.2	2005	0
2001	0	2006	0
2002	0	2007	0

fishery resource. Data on fish catch is given in Table 1 for the period 1960–2006 in the Republic of Karakalpakstan.

In 1957–1958, more than one million units of valuable muskrat skins were obtained from lakes in the lower Amudarya. However, after the critical low water years of 2000–2001, their population has completely disappeared. Table 2 shows muskrat skin production in Karakalpakstan.

The coastal zone has lost its vital attractiveness and landscape value, due to decreasing water volume in the lower reach of the river. By now, the area of lakes in the Amudarya delta has declined from 300,000–40,000 ha. Moreover, mineralization of water has sharply increased and in some places, water contains up to 60 g/L of dissolved solids (Lake Agushpa) (Irrigation management for addressing the desertification processes in the Aral Sea basin, Tashkent, 2005).

Increase in the Area of the Dried Bottom of the Sea

As the natural landscape zone of the Amudarya delta with its tugai flora and specific wildlife degraded, wildlife numbers fell sharply. The salt-dust from the dried sea bottom together with the declining sea level resulted in desertification and a decrease in natural bonitet of lands in large territories. The reed areas that were used as the main pasture areas for livestock were sharply reduced (30-fold). The intensive process of drying up of the sea initially took place in the sea bays Jiltirbas and Adjibay. The speed of coastline regression in Jiltirbas bay zone was 2.7–3 km annually. The intensive drying process corresponded to the period 1980–1985 and its area accounted for 2,387 km^2 (in the south). According to V.A. Dukhovny (V.A. Dukhovny et al. 2003) total direct and indirect socio-economic losses from the ecological disaster in Priaralie are estimated at 144.8 million USD.

Assessment of Loss in Agricultural Production

From 1965 to 1996, the total area of irrigated lands increased from 220,000–500,000 ha, while rice crop area increased from 5,400–110,000 ha in the Republic of Karakalpakstan. During some years, 448,500 t of cotton and 368,300 t of rice were produced. In low water years a 3.5, 3 and 30-fold decrease was observed in the areas of irrigated land, cotton and rice respectively (Fig. 3).

The value of gross yield of crops depends on water availability. The production of the highest and most stable yields correspond to the period from 1983 to 1993, when water availability in the river was enough to meet crop water requirements. During this period, the average values of gross cotton and rice yields amounted to:

Cotton: 337.500 ton
Rice: 301.000 ton

Meanwhile in low water years, the shortfall in yield fell to 250,000 t (Fig. 4).

For ecological objects and the agricultural sector only, the losses caused by water shortages amount to 219.8 million USD in low-water years, 164.3 million USD during average water availability years and 144.8 million USD in high-water years. For mitigating unfavorable ecological conditions and the socio-economic situation in the lower reaches of Amudarya, it is necessary to make decisions at a regional level. The main principles are:

- Development of a long-term concerted policy for Central Asian countries on water supply to the Priaralie region, on the principle of existing international counterparts.

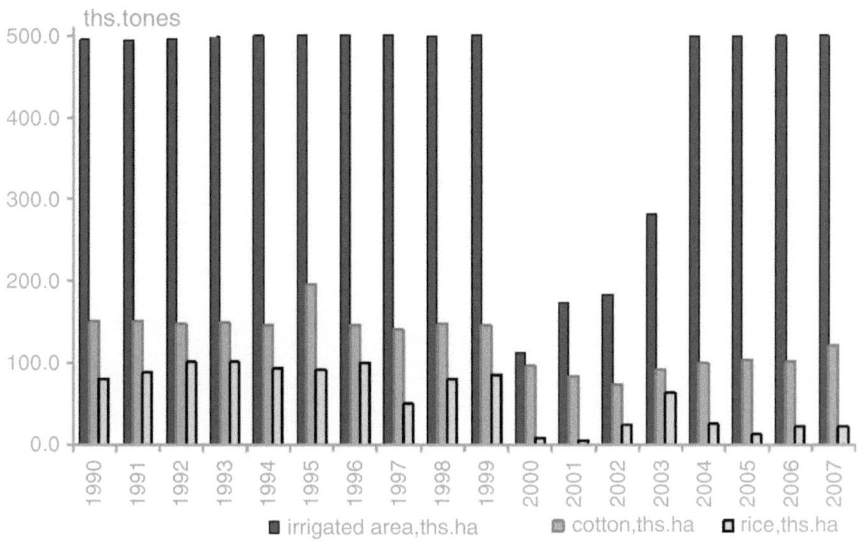

Fig. 3 The dynamics of irrigated lands in the Republic of Karakalpakstan, including cotton and rice fields for the period 1990–2007

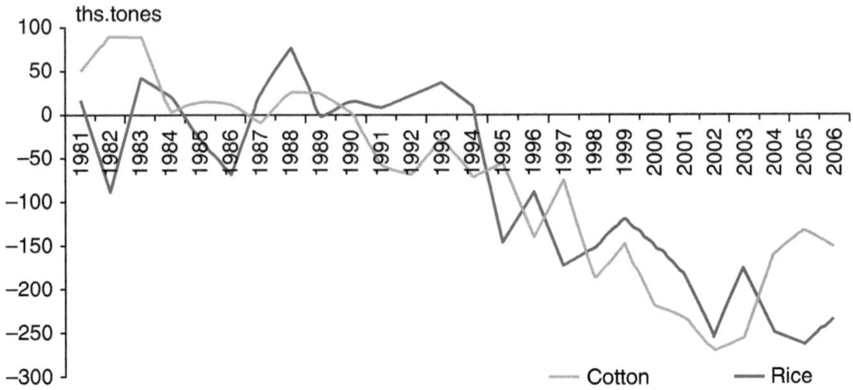

Fig. 4 Losses in the gross yields of cotton and rice in the Republic of Karakalpakstan for the period 1981–2006

- Increasing the role and authority of BWO « Amudarya » by transferring all water withdrawal objects, large water works and reservoirs to the balance sheet of this organization so it can ensure a guaranteed water supply to Priaralie region.
- Further increase in agricultural production in the Aral Sea basin should be achieved by increasing the efficiency of irrigated hectare, rather than land reclamation.

References

V.A. Dukhovny, Joop De Shutter – « South Priaralie – new perspectives», «Science for the world » Project, Tashkent, 2003.

Experience in Controlling Structures of Volzhski-Kamsk Cascade of Reservoirs and Issues of Socio-Economic and Environmental Stability in the Volga Region

Sergei Yevstaphievich Bednaruk

The Volga basin, with a catchment area of more than 1.35 million km^2, is the largest river basin in Europe and among the biggest basins in the world. The mean annual flow of the Volga River is about 260 km^3. More than 60 million people live in the basin area. As a result of the construction of 11 hydro schemes with reservoirs on the Volga River and its main tributary Kama, between 1930 and 1970, one of the world's largest water-management systems was formed, with a total reservoir volume of about 190 km^3 and a net storage capacity of 80 km^3. The established system has a number of functions, including electricity generation, large-capacity navigation, water supply to the population, and irrigation.

Construction of hydro schemes and reservoirs, especially in the Lower Volga (Kuibyshev and Volgogradsky) radically changed the hydrological regime of the Volga River by redistributing its stream flow from high-water spring to low-water summer and winter. In addition, the hydro schemes blocked the passage of valuable fish species (such as the unique Caspian sturgeon) to spawning places. By the time the final Volgogradski hydro project within the cascade was put into operation (late 1950s), it was evident that without special measures the natural and economic systems in Lower Volga that depend completely on spring flooding of Volgo-Akhtubinskaya floodplain and Volga delta, were doomed to degradation and disappearance. To conserve natural ecosystems and the traditional living conditions for residents of the Lower Volga, the Government of the USSR decided in 1959 to make annual, specific spring water releases through the Volgogradski hydro scheme into the lower reaches of the Volga River to benefit fisheries and agriculture, as a "temporal measure to finish fishery and agricultural reconstruction in Lower Volga". Since measures planned for the Lower Volga were not implemented, those specific spring releases have been kept until now.

S.Y. Bednaruk (✉)
Ministry of Natural Resources of the Russian Federation, Center of Russian Waterworks Inventory and State Water Cadastre, 2/ 2-a, Pryanishnikov Str., Moscow, 127550, Russian Federation
e-mail: bse@waterinfo.net

C. Madramootoo and V.A. Dukhovny (eds.), *Water and Food Security in Central Asia*, NATO Science for Peace and Security Series C: Environmental Security,
DOI 10.1007/978-90-481-9974-7_15, © Springer Science+Business Media B.V. 2011

Great changes have taken place in the economic activities of the basin over the last 50 years. The state cascade management system was established and still functions with involvement of multiple concerned departments and organizations. During the spring period, hydro schemes are operated in special environment-oriented modes.

There are a number of different problems affecting management decision-making, including development of land in downstream and on upstream sides of reservoirs, poor condition of engineering infrastructures at public utility companies and agricultural equipment, the incomplete cascade, respective navigation problems, etc.

Inspection of the cascade operation was made in 2005–2007, during which time a number of severe or extreme events occurred:

- 2005 – high water – problems related to water passing through the cascade's hydro schemes and flood prevention
- 2006 – extremely low spring tide levels and summer drought – problems related to special spring pass in Lower Volga, very tense environmental situation, growing social tension
- 2006–2007 (winter) – historical maximum inflow to reservoirs of the cascade; problems related to high-water by-pass in winter period
- 2007 (spring) – fulfilling of optimal environmental releases to the lower reaches of Volga under low spring tide

Recently, despite the smoothly running cascade management system, contradictions have arisen primarily between territorial interests (and respective entities of the Russian Federation) connected with both the socio-political and economic changes in the country and the hydrological changes over the last 15 years.

A description of current cascade management system is given below. Figure 1 shows the general pattern of Volzhski-Kamsk reservoir cascade management system. The management objects are the largest hydro schemes in the Volga basin as shown in the Figure in the left upper corner.

The management system consists of the following blocks:

- Receiving and processing on-line information
- Drawing up proposals on hydro scheme operation regimes (information-analytical center, which is FSUE "Registry and Cadastre Center")
- Representatives of concerned departments and organizations related to the water use through the cascade of reservoirs and the protection from negative water effects in the reservoirs' affected zone
- Management decision-making (inter-departmental task force (ITF) and decision makers)

In the block for receiving and processing on-line information, the main information providers are:

- Public corporation "System operation" Central Dispatch Administration of the Unified Energy System of Russia (UES CDA) – data on operation modes of hydropower station hydro schemes (excluding such HEPS hydro schemes

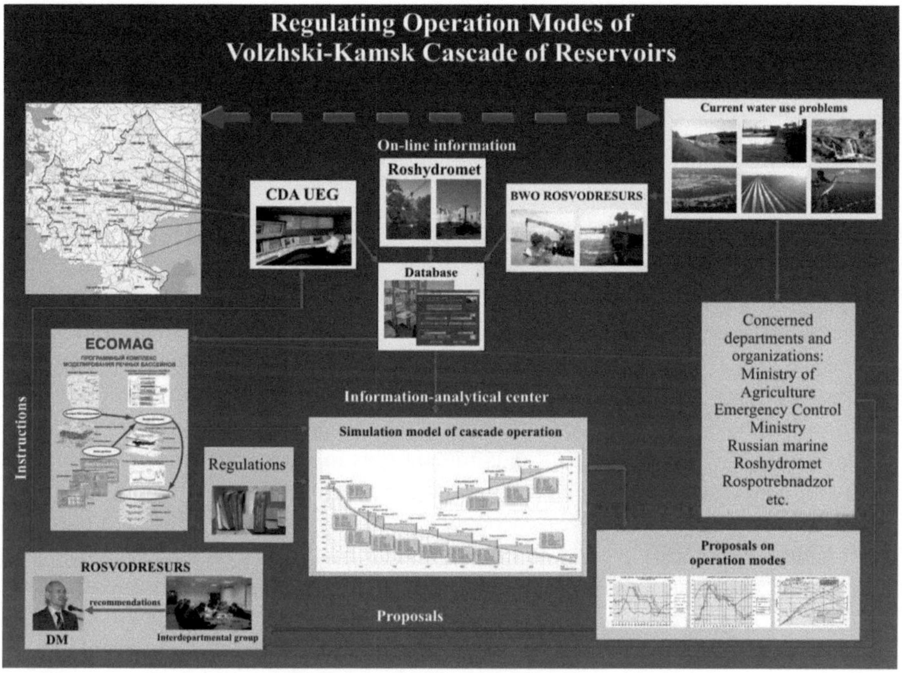

Fig. 1 Regulation of water regime of Volzhski-Kamsk Cascade of Reservoirs

as Sheksninsky and Ivankovsky hydro schemes belonging to non-energy organizations)

- Hydrometeorological center of the Russian Federation (Institute of Rosgidromet) – data on: water levels in reservoirs and rivers; inflow to reservoirs, water temperature in certain gauging stations; weather data including air temperature and humidity, rainfall, atmospheric pressure from all weather stations of the Volga basin and adjacent areas (in total, more than 700 stations); soil moisture and frost zone from basin's agrometeostations; water storage in snow; monthly and quarterly forecasts of inflow to reservoirs; forecasts of ice development on basin's rivers and reservoirs
- Federal state unitary enterprise "Moscow Canal" at Rosmorrechflot of Russia's Ministry of Transport – data on operation modes of Ivankovskoye and Verkhnevolzhsky reservoirs on the Volga River
- Basin water authorities at the Federal Agency for Water Resources (Moscowsko-Oksky, Verkhnevolzhsky, Kamsky and Nizhnevolzhsky) in their jurisdiction areas – data on technical and environmental conditions of reservoirs and river reaches; current problems and requirements of water users, executive agencies and local authorities

Data from UES CDA are transmitted daily through email to Rosvodresursy (Federal agency for water resources) in the form of a DBF-file and stored on a network drive of a local computer network of FSUE "Registry and Cadastre Center", to which one of the computers from the on-line regulation division is connected. The data are used for keeping an on-line e-journal on the operation modes of Russia's reservoirs, which is a database of the large reservoir's hydro scheme performance measures.

Data from the hydrometeorological center of the Russian Federation are transmitted by internet between local computer networks of the Center and the FSUE "Registry and Cadastre Center" immediately after their acquisition from territorial branches of Rosgidromet, in coded data formats established by the latter. After decoding, these data are placed in respective databases of FSUE "Registry and Cadastre Center" for further use, including information support of water management in Volzhsky-Kamsk cascade of reservoirs. In addition, data on water levels for the most recent 10 days from all gauging stations in Russia are posted on FSUE's web-site (www.waterinfo.ru) under the project "Russia's rivers".

Every day, data from FSUE "Moscow Canal" are transmitted through phone to the on-line regulation division of Rosvodresursy and then to FSUE "Registry and Cadastre Center" and inputted manually into an e-journal on the operation regimes of Russia's reservoirs.

Data from basin water authorities are transmitted through various communication facilities (phone, fax, e-mail, cable, ordinary mail) to the on-line regulation division of Rosvodresursy and, if necessary, to "Registry and Cadastre Center" for reservoir operation variants calculation as additional limitations.

All received data are used in FSUE "Registry and Cadastre Center", which acts as an information-computer center for calculations and development of proposals on reservoir operation modes.

Schematically, the process of the long-term planning of Volzhsky-Kamsk hydroscheme operation modes and the development of proposals on hydroscheme regulation consists of two main stages:

1. Based on scenario calculations by flow formation model, dynamics of lateral inflow to Volzhsky-Kamsk cascade's reservoirs is estimated
2. Based on scenario simulations of lateral inflow by simulation model of reservoir cascade functioning, operation modes of Volzhsky-Kamsk cascade's hydroschemes are calculated

The information-modeling system (IMS) ECOMAG (ECOlogical Model for Applied Geophysics) is used as a platform for calculations of flow formation and lateral inflow to reservoirs in the "Registry and Cadastre Center". The system includes: the mathematical model ECOMAG, a specialized geographical information system (GIS), a database on terrain characteristics and control frame.

The ECOMAG system allows the calculation of hydrological characteristics (including inflow reservoirs) from weather data and creation of a hydrological picture (mapping description) of terrain, including diagnostic, prognostic and simulation hydrological maps and river basin pollution maps under different

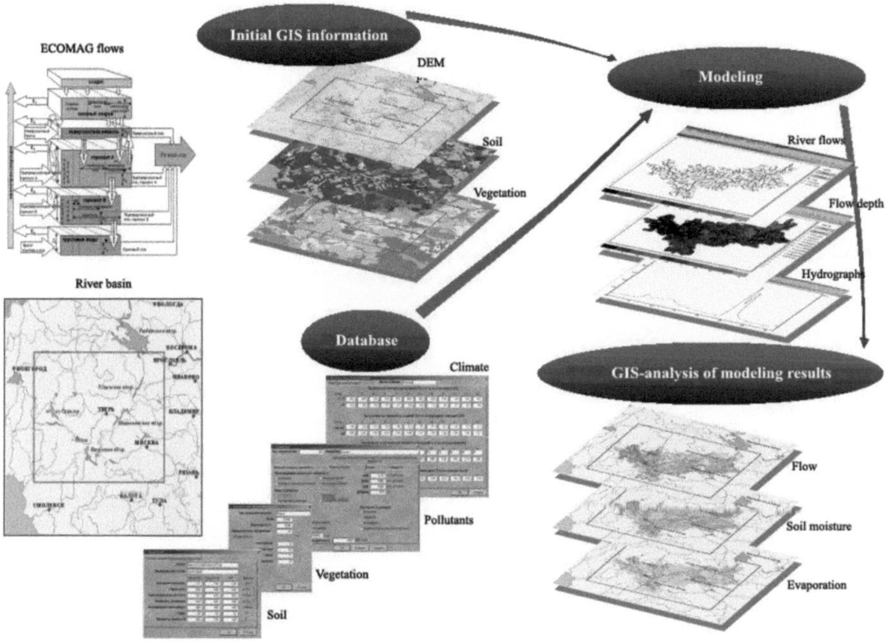

Fig. 2 The general structure of the ECOMAG modeling system

hydrometeorological conditions and anthropogenic pressure. The general structure of IMS ECOMAG is shown in Fig. 2.

Schematization of the catchment area and river network in the Volga basin is made on the basis of thematic maps (1:1,000,000), using special GIS-based software ArcView. To this end, the GIS database contains a number of layers of the basin's thematic e-maps (digital elevation models – DEM, hydrographic, soil, landscape and other maps, Fig. 3).

These layers were rastered on square net (2 × 2 km), and each cell was attributed a respective value of elevation, number of soil type and landscape. Using special GIS software, the tree-like model structure of the river network was constructed and watershed lines, boundaries of individual inflow catchments, were identified. At the next stage, the relief work grid was corrected to define more exactly the model river network in those places where it does not correspond to the actual (vector) one. Information on characteristics of slope elements and river network structure is then transmitted to IMS « ECOMAG » where from respective databases of each element are attributed to model parameters (soil, vegetation, etc.).

An example of model Volga river network is shown in Fig. 4.

Figure 5 shows catchment areas of the reservoirs of Volzhsky-Kamsk cascade and simulation elements of catchment areas received by special DEM-based GIS system.

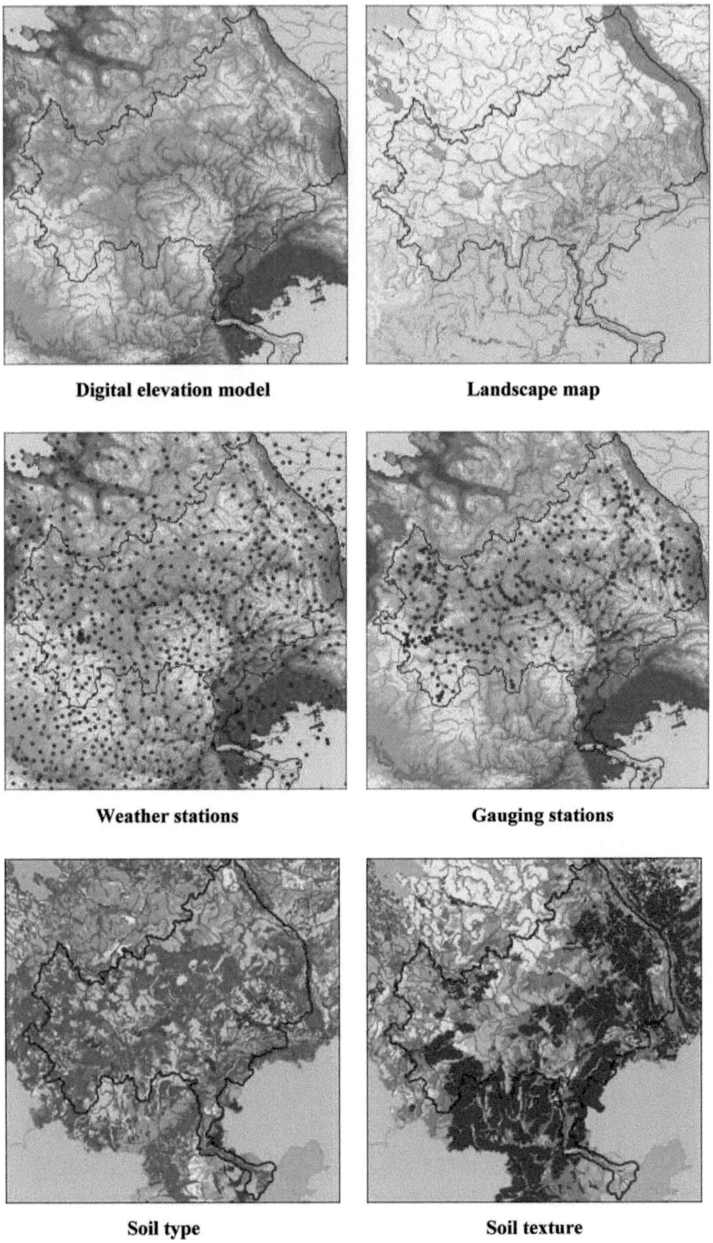

Fig. 3 Basic GIS-information

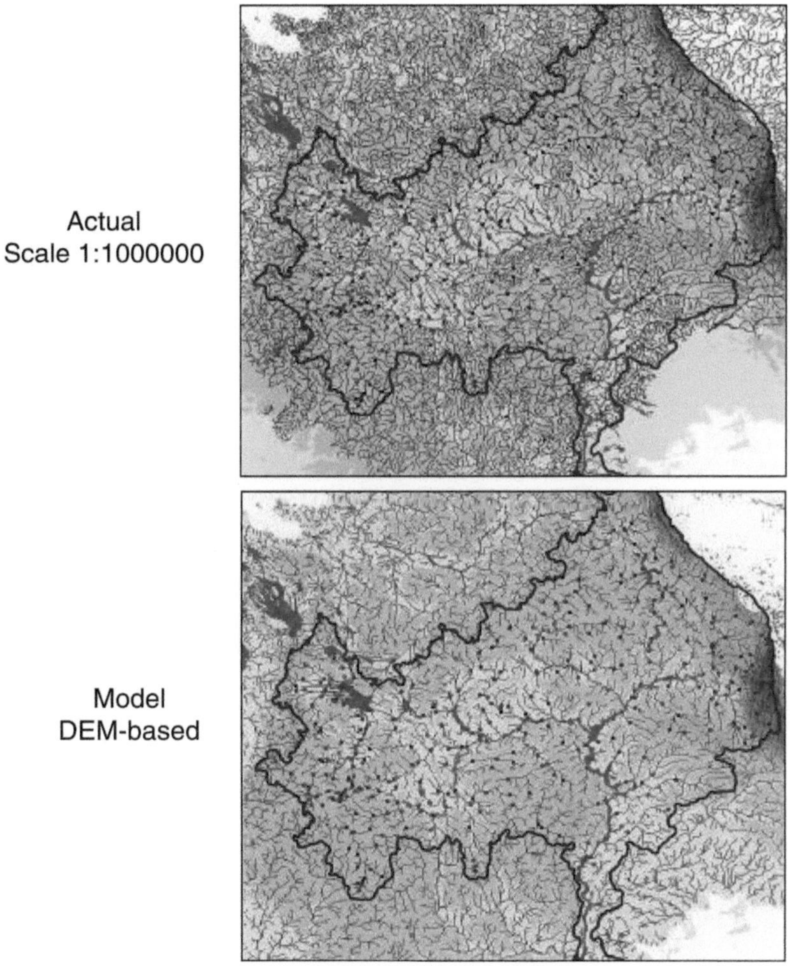

Fig. 4 Structure of actual (scale 1:1,000,000) and model (DEM-based) river network in the Volga basin

Scenario computations of inflow to the reservoirs of the Volzhsky-Kamsk cascade using the ECOMAG model consist of the following stages:

1. The model makes computations using weather information for current date of scenario generation for no later than previous yearly period.
2. In order to estimate the processes of the hydrological cycle and lateral inflow to reservoirs to the next period (for example, the second half of 2007), scenarios of meteorological process development are set for this period. Since weather forecasts with daily resolution are not available for such periods, the model sets the following meteorological scenarios for the second quarter: it is assumed that weather from June 30 to December 31 in 2007 would be the same as during

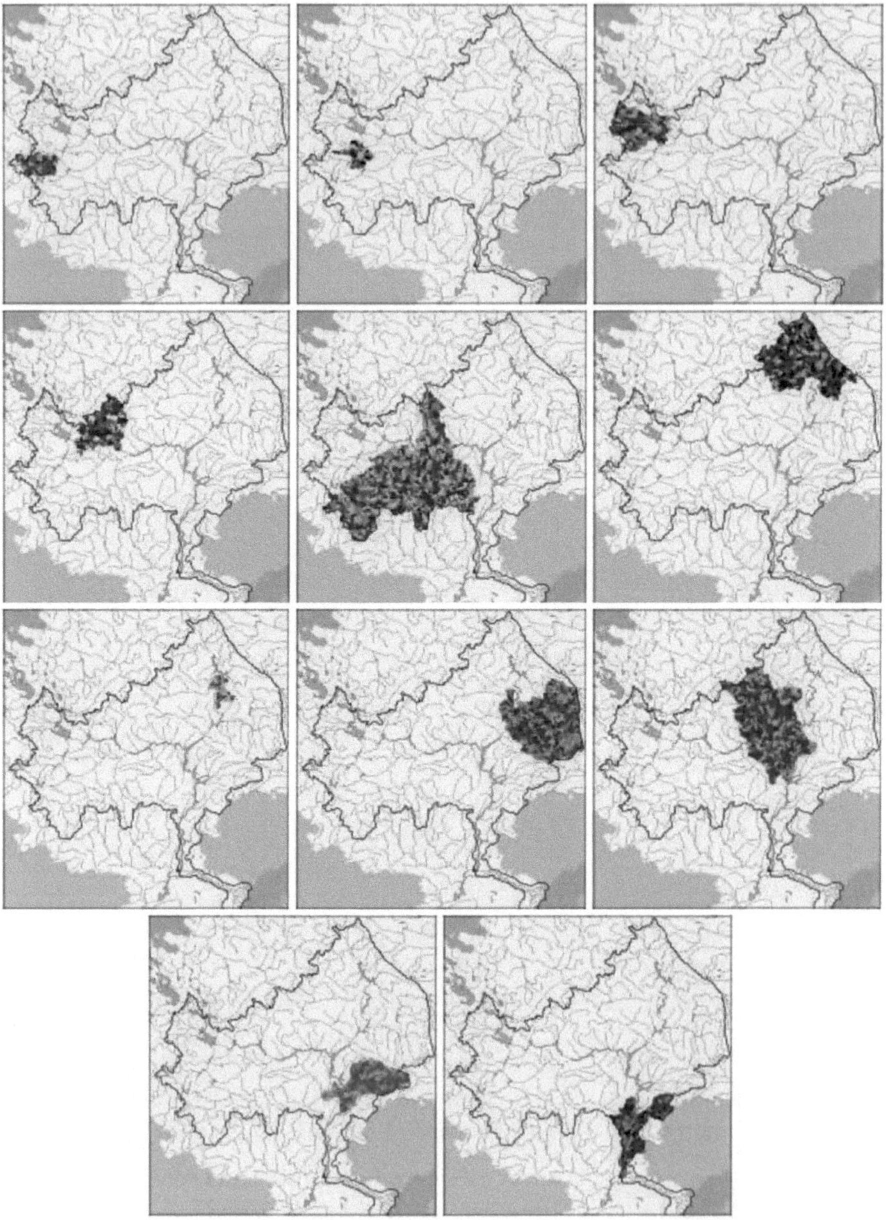

Fig. 5 Catchment areas of Ivankovsky, Uglichsky, Rybinsky, Gorkovsky, Cheboksarsky, Kamsky, Votkinsky, Nizhnekamsky, Kuibyshev, Saratov and Volgogradsky reservoirs (With selection of elementary catchments)

the same dates in 2005 or in 2006 – according to another scenario, or the same as in 2003, etc., using historical weather information for previous years. Besides, for the nearest 10 days since the date of the beginning of scenario computations (i.e. since July 1 to July 10), the model considers the weather forecast of Rosgidromet for the largest settlements located close to the reservoirs, which is taken from the internet. For this period of time, a synthetic weather scenario is built by choosing similar forecast days from the previous period. In addition to the above-mentioned meteorological scenarios of weather development, we use a synthetic weather scenario, which is generated by averaging weather characteristics by days in long-term dimension, using the archived weather database.

The hydrological model ECOMAG then computes, for the period of meteorological scenarios, the flow formation processes in reservoir catchment areas and the scenarios of lateral inflow to them. The outputs of ECOMAG are scenario hydrographs of lateral inflow to the reservoirs of Volzhsky-Kamsk cascade for the calculation period. In addition, a synthetic inflow hydrograph is built by averaging all scenario hydrographs by date. Those scenario computations are used for the development of operation modes of hydro schemes when regulating the reservoirs of Volzhsky-Kamsk cascade.

Planning of the operation modes of hydro schemes located in the Volzhsky-Kamsk cascade, including by-pass of spring floods, is made on the basis of multivariant calculations of the operation modes for the whole range of probable hydrological conditions (as predicted by Rosgidromet and scenario ones by ECOMAG), using the simulation model VOLPOW. Implementation of the mathematical simulation model VOLPOW connects the hydraulic calculations of water flow in reservoirs with the water balance calculations for the cascade of hydro schemes and, hence, with water use management. The hydrological block of the model is based on unsteady flow equations. The water-management block includes requirements and limitations contained in dispatch rules of hydroscheme operation, that are the relationships between the required (and allowable) releases of water through a hydroscheme, the water level in upstream pool and the time of year. Connection of hydraulic calculations with the water-management block is made by imposing for the unsteady flow equations of boundary conditions resulted from dispatch regulation rules for hydroscheme operation and from structures' flow capacities.

Generally, in order to solve a given problem, detailed information about width of control stations, slopes, roughness coefficients, etc. is needed. Usually, such detailed information is not available in engineering practice. Often there is some integrated information in the form of curves of reservoir volumes, relationships between discharge and level in control stations, nomograms of flow capacity, etc. Therefore, finite-difference algorithms are modified in the VOLPOW model by integrating some terms into certain functional relations that allow using detailed and integrated input morphometrical and hydraulic information.

In the schematization of reservoirs of the Volzhsky-Kamsk cascade, each reservoir is divided into a few calculation sections (from 1 to 4). Available integrated and point morphometric and hydraulic characteristics are used for each section.

Boundary conditions are formulated in section lines of hydro schemes. These conditions are determined by flow capacities of the structures and dispatch rules of hydroscheme operation.

Input hydrological information is a calendar series of lateral water inflow to calculation sections (in particular, calculated by ECOMAG). Simulation calculations can be made with the time resolution from 1 day to 1 month (experience showed that for the Volzhsky-Kamsk cascade the following calculation time steps are enough: flood period – 1–10 days; low-flow period – 10 days to 1 month). The output of the simulation model VOLPOW are water balances for each reservoirs in the cascade and calculated regimes of discharge and level in section line for lateral inflow scenarios.

The long-term planning of the operation modes of hydro schemes in the Volzhsky-Kamsk cascade is made on the basis of lateral inflow scenarios and calculated scenarios of hydroscheme operation modes. Among the set of calculated scenarios, the most plausible scenario of lateral inflow to the reservoirs is selected, as well as schedules of hydro scheme operation within with cascade, with account of optimal regime of reservoirs, their complete filling and provision of effective annual special spring by-pass through the Volgogradsky hydro scheme to the Volga downstream for the benefit of fisheries and agriculture. Selection of the most plausible scenario of lateral inflow is made on the basis of current hydrometeorological situation in the Volga basin and Rosgidromet's forecasts, using total volumes of lateral inflow to reservoirs over the current quarter and its first month. Moreover, calculations on meteorological scenarios on the date of calculation that do not correspond to current weather are rejected at once.

From the set of lateral inflow scenarios calculated by ECOMAG, the more negative maximum and minimum inflow scenarios are selected and scenarios of optimal hydroscheme operation modes calculated by the simulation model VOLPOW on the basis of selected inflow scenarios are then chosen. The results of lateral inflow scenarios and of the consequent scenarios of optimal hydroscheme operation modes for the most plausible maximum and minimum inflow scenarios are disseminated among concerned departments of Rosvodresursy and presented at meetings of the Inter-departmental Task-force for Regulation of the Volzhsky-Kamsk cascade reservoirs in the form of proposals for decision-making by the authorized federal executive agency (Rosvodresursy).

During the low-flow period, calculations are made every 10 days, while in spring tides, scenario calculations by the simulation models are made once every 5 days, taking into account current hydrometeorological and water-management conditions and weather forecasts for nearest 10 days.

Actual useful inflow is calculated for the previous period at each stage (the so called "inverse problem"). Based on calculation results, the expected design hydrographs of inflow to reservoirs are changed for the next period.

At the end of each calendar quarter, by comparing scenario calculations with actual data on inflows to and releases from the reservoirs, an analysis of the efficiency of the long-term hydroscheme operation planning is made, causes of deviations are identified, and proposals for the improvement of the Volzhsky-Kamsk cascade situation management system are formulated. An important element of the described management system is the Inter-departmental Task-force Group (ITG), which is formed among the representatives of concerned federal ministries and departments, major water users, executive agencies, research and production organizations. ITG is established according to an Order of Rosvodresurcy and acts on the basis of regulations approved by Rosvodresurcy as well.

The main objective of ITG is to create conditions for on-the-fly provision of demands of all actors interested in hydroscheme operation modes under current hydrological and water-management conditions. ITG includes representatives of Rosvodresursy, Ministry of Environment, Ministry of Agriculture, the Russian Emergencies Ministry, Rosgidromet, the Russian Marine and Inland Water Transport (Rosmorrechflot), Rosstroy, Rospotrebnadzor, Rosselphoznadzor, Administrations of Astrakhan and Volgograd provinces, Public Corporation "RAO Unified Energy System of Russia", Public Corporation "GidroOGK", Public Corporation "RAO CDA UES", FSUE "Caspian fisheries research institute", FSUE "Registry and Cadastre Center". In addition, members of the Federation Council or delegates of the State Duma, representatives of legislative and executive agencies, local authorities, basin water administrations at Rosvodresursy, basin waterway administrations of Rosmorrechflot, shipping, building and other companies, and research and community organizations take part in the meetings of the Task Force by invitation or on their own initiative.

Based on the developed proposals for operation modes of the hydro schemes (prepared by FSUE "Registry and Cadastre Center") and the current hydrological conditions (reported by representative of Rosgidromet), ITG meetings discuss and develop recommendations regarding the operation of the cascade's hydro schemes for the next period. The recommendations are documented in the minutes of Task-force meeting, major points of speeches and opinions of ITG members and participants are also noted.

It should be noted that the ITG recommendations form the basis for decision making but are not obligatory for implementation since authority to make decisions on regimes of reservoir filling and drawdown is given to the Federal Agency for Water Resources (Rosvodresursy). Thus, responsibility for decisions made is fixed clearly. Within Rosvodresursy, regulations set decision makers, including the Director of Agency, one of his deputies and the chief of real-time regulation. Moreover, lower-level DM can make decisions only if the upper-level DM is absent. In most cases however, the DM follows the recommendations of ITG, and in only a few cases were decisions in contradiction to the recommendations. This is the case, when momentary interests of most bodies and organizations are the same but contradict the goals of guaranteed water supply and security as specified in the effective regulatory legal acts. Thus, this causes a risk of breaking guaranteed water supply or emergencies as a result of water-logging or flooding of land and settlements.

Decisions made by DM are documented in the form of directions signed by DM, sealed by Rosvodresursy and submitted immediately by fax to the system operator of CDA UES. On the basis of the decisions received, the operator develops an hourly operation mode of the cascade's hydro schemes and submits it to HEPS management for implementation. Thus, this closes the control cycle of the cascade's hydro schemes.

Alternative Sources of Water Supply to Minor Water Consumers in the Deserts of Central Asia

Agajan Geldiyevich Babaev

Central Asia is a united natural and social-economic complex with an area of about 4 million km^2, and a population of over 57 million people. In terms of administrative divisions, there are five independent countries located on its territory: Kazakhstan, Kyrgyzstan, Tajikistan, Turkmenistan, and Uzbekistan. Being in the buffer zone between East European and West Asian countries, it has closely, in historical terms, experienced their political, economical, and cultural influence in the course of its own development. The Great Silk Road ran through this region and played the role of a connecting link.

In terms of hypsometry, this vast area ranges from 132 m below sea level (Karaghie depression, Western Kazakhstan) to 7,495 m above sea level (Ismoil Somoni Peak of Tien Shan, Tajikistan). Here, landscapes and climatic zones alternate, different in their origin and natural structure. The largest part of the area is occupied by deserts and semideserts. The predominance of arid areas is explained by the closed inland geographical position of the region, where biological and landscape diversity formed under continental climate conditions with limited atmospheric precipitation. In terms of litho-edaphic arid conditions, sandy, clay, loessial, stony, and salt deserts are observed in Central Asia. Clay deserts are of particular interest from the standpoint of water supply and occupy a total area in Central Asia of roughly 12 million ha, which is 1.5 times the total area of the currently irrigated lands in the region.

The Central Asian region, possessing great natural and human resources as well as high economic potential, attracts a great deal of attention from a number of developed countries. After countries in the region became independent, industry and agriculture began to develop and the GDP increased significantly. All of these achievements have been made thanks to water resources. Since the dawn of time water had, and has, a special place under the arid conditions of Central Asia. It has an extremely high value here. It is no coincidence that both among past and present

---

A.G. Babaev (✉)
Interstate Commission for Sustainable Development of Turkmenistan,
142 Hero of Turkmenistan, A. Niyazov Str., Ashgabat, 744000, Turkmenistan
e-mail: babaev@dcoret.ashnabad.su

C. Madramootoo and V.A. Dukhovny (eds.), *Water and Food Security in Central Asia*,
NATO Science for Peace and Security Series C: Environmental Security,
DOI 10.1007/978-90-481-9974-7_16,

generations the following proverbs existed: “A drop of water is a grain of gold” and “Where water is absent, land is absent”. Over a long historical period the deficit of fresh water for drinking, irrigation, and stock watering made the local people invent various techniques and technologies. They constructed primitive but wise water-catchment and water-saving facilities: dams; reservoirs; channels-aryks (irrigation canals); kahrezs (drainage galleries); sardobs (dew mounds); khaks; khauzs (ponds); wells; etc. The people could find fresh water by intuition and, using limited water reserves, they supported their household needs, watering of sheep and camels in void pastures and in some areas, they kept small-oasis farms on patches of land.

The traces of ancient cattle-breeders and farmers can still be found and are evidence of the hard labour and creative talents of desert dwellers. Recent scientific discoveries and results of archeological excavations have shown that as far back as the late Stone Age and Bronze Age small, oasis agriculture was developed in Southern Turkmenistan. Grain crops (wheat, barley, millet, etc.) used to be cultivated and watering was carried out by means of primitive irrigation technology.

In this age of scientific and technological advances, old approaches and ways of supplying water to minor consumers have undergone drastic changes. Great dams and reservoirs have been built, long-stage channels constructed; large water-desalinating plants are operating; and the level of technical equipment of irrigated agriculture is manifold. However, these water-related engineering achievements in oases have barely been applied with regard to the problems of water supply to small cattle-breeding settlements in deserts proper. The construction of special, kilometers-long conduits from rivers and channels or water delivery by tank lorries to remote cattle farms turned out to be unprofitable from an economic point of view. Alternative sources of water supply to small water consumers in the desert are still the most profitable.

As is known, huge groundwater reserves in the Central Asian deserts are generally highly mineralized and are unusable. At the same time, fresh water lenses, floating on the salt groundwater, have been discovered here and thoroughly investigated over the last decades. The areas of their spread, with the thickness of the water-bearing horizon of 100–150 m, vary from a few hundred to a few thousand km^2. There are several views concerning the origin of these fresh water lenses. One group of scientists thinks that they have formed because of rainwater flowing down from bajada surfaces and following immersion of those under hilly-ridgy sands to the level of salt groundwater; another group associates their origin with the age-old water filtration from rivers and freshwater lakes; and a third team of scientists considers the lenses to have relict origin.

In olden times, eight such freshwater lenses, having total static reserves of over 60 km^3, were discovered and thoroughly investigated in the Karakum desert: Yaskhan (10 km^3); Cherkezlin (2 km^3); Balkuin (0.45 km^3); Eastern-Zaunguz (3.4 km^3); Djillikum (8.40 km^3); Repetek (0.84 km^3); Karabil (25 km^3); and Badkhyz (19 km^3). For example, the Yaskhan lense, is located in the Western Karakum area, in the dry bed of the Uzboy River that flows down from Sarykamysh Lake, where part of Praamudarian water drained during the Upper Quaternary period. In the 1960s, it became the object of diversified investigations with regard to water supply. At the Yaskhan lense, a unique water intake operation technology was developed, which not only reduced the critical water shortage in the pasture area, but also improved, to some extent, the water supply to the large Nebitdag industrial center by constructing a water conduit 142 km long with overall water intake over 4,000 m^3/s.

The kariz system is referred to as an alternative source of water supply to small settlements. Generally, kariz technology has been developed for conditions of arid sloping submontane plains where the precipitation is about 200–300 mm/year. The filtration water, forming layers of groundwater at different levels of proluvial deposits, comes to the surface in the form of low-yield springs and small episodic rivers. In order to withdraw some part of the fresh underground water, the local people invented a peculiar technology for the construction of the kariz system. Called by different names, the kariz system exists in many countries of the arid zone: Turkmenistan, Uzbekistan, Algeria, Yemen, Morocco, China, and others. For example, in Turkmenistan, on the submontane plain near Kopetdag, a few operating kariz systems constructed by our ancestors still remain intact. The kariz construction technology is quite simple. Along the slope of the relief a subsurface gallery is constructed downward from the water-bearing horizon with the following dimensions: 3–5 km length; 1–1.5 m width; 1.3–1.5 m height; and has 50–60 vertical wells that are spaced 30–100 m apart. The level of the wells toward the water-bearing horizon is within 40–50 m from the kariz source and 2–3 m from the water appearance at the surface. At the construction of the kariz, those wells serve to carry water over the ground from the gallery, and later as a hole with the purpose of observation. The volume of free-flow kariz water carried out to the surface amounts to 20–50 L/s. Part of the kariz water was used by people for irrigation of their plots of cereal crops, vegetables, melons and gourds, and vine. As for its quality, kariz water is perfectly pure and fully meets the highest world standards. In Turkmenistan, in the 1950s, operating karizs numbered more than 200 with a total yield 2,500 L/s. In the 1970s, 103 remained; and nowadays, there are only 38 units. Keeping such water facilities in good order in an arid zone required much ingenuity and manual labour during constructing and operation and is evidence of the great services of small teams of enthusiastic people. Unfortunately, at present the kariz system of water supply has dwindled.

Among local water supply sources is also the rainwater that periodically accumulates at the connection of sloping submontane plain and sand-clay desert. Temporary land runoff, formed under continuous precipitation over 10–15 mm, makes a wetted belt with relatively loose sierozem-meadow ground and rare herbaceous and shrubby vegetation. Channel-shaped depressions with areas of 0.5–1.0 ha, called by the local population "oytak", have a high moisture content. They are long used for small oasis (oytak-based) farming. Oytak farming resembles in many respects rain-fed farming, the development of which does not require much material and labour costs, since water arrives spontaneously.

A certain amount of fresh water can also be obtained by using solar energy, an inexhaustible resource in the desert. In the 1980s, the Solar Energy Institute of the

Academy of Sciences of Turkmenistan constructed a greenhouse-type solar desalination system with a capacity for drinking water of 600 m^3 per year. The system operated for 5 years, but was eliminated because of its low efficiency and relatively high prime cost of the water desalination.

The Institute of Deserts of the Academy of Sciences of Turkmenistan has carried out experimental work in the Zaunghuz Karakum area concerning salt groundwater desalination by means of the natural cold of the winter period. A concreted (asphalted) channel-shaped site, 200 × 200 m was built in swale features and filled with mineralized underground water of 20 g/L, in 10–12 cm layers. The upper 2-cm water layer is exposed to freezing overnight and then the bottom, non-frozen part of the salt water is poured out through special holes, and the ice crust gravitates to the bottom of the site. This operation can be done many times. As a result, a multilayer ice stratum arises resembling a layer cake; as warm days come, this stratum gradually thaws and flows down to a special water-accumulating reservoir. This technology can successfully be applied in arid areas of the northern deserts, where low air temperature is observed over winter.

Temporary rainfall runoff, formed at the surface of clay deposits, is the most reliable and economic source of water supply to minor consumers in deserts. Accumulation is due to the peculiar condition of atmospheric precipitation characterized by excessive variability in time and great unevenness of distribution over the area. Precipitation during the autumn–spring period year is observed and may vary from 10 mm per one rain to 70 mm. Water volumes of over hundreds and thousands of cubic meters form on the clayey (takyr) surfaces after heavy rains.

Takyr is one of typical types of clay desert landscapes. High concentrations of alphitite (up to 75% and over), form extremely low water-permeable surfaces with slight slope (0.001%), and takyrs and takyr-like soils serve as excellent catchment areas. Ideal flatness and polygonal fissuring is characteristic of these. Takyrs and takyr-lke lands occupy vast areas in the deserts of Central Asia. For ages, they have been used as catchment, and the rainfall runoff accumulated served as the only water supply source in the desert. Through the centuries, the local people of the deserts have worked out a number of original technologies for the collection and keeping of the takyr surface flow. They dug shallow (1–2 m) open holes – khaks – in which rainwater accumulated and where the flow was quickly filtrating into the soil, they constructed self-flowing wells called "chirle". Over the long term, the collection of precipitation and storage technology have been improved. Engineering facilities with simple construction – covered storage of the takyr precipitation water (sardobs) – have remained to the present day. Such water reservoirs used to be built near to caravansaries and active roads and provided travelers and their cattle with fresh water for dry periods of almost one year.

Research work studying takyrs and takyr flows allowed the detection of the mechanism of their formation, factors influencing water volume, as well as developed flow calculation methods. Precipitation depth per one rain and the area of takyr surface are the dominant factors that determine the amount of the flow. For Central Asia, the potential resources of the takyr flow average 700 million m^3 per year, and for Turkmenistan just over 300 million m^3 per year. In the Central Karakum area, where the average annual rainfall is about 100 mm, takyr flow from a 1 km^2 area comes to 15,000 m^3 per year with mean water content. By accumulating and keeping even 10% of the local surface flow, one can meet the water requirements of one cattle-breeding farm.

Currently only 900 khaks, 350 sardobs, and 500 self-flowing wells-chirle have been constructed and are operating. To improve the traditional folk methods of local flow water collection and enhance their efficiencies, the Institute of Deserts of the Academy of Sciences of Turkmenistan established a research base in the Central Karakum area in 1965. Here, in the hole Karrykul, on the takyr with an area of 250 ha, a hydraulic complex has been built and improved by new engineering and design solutions. It consists of a waterworks (3-m deep pool and an area of 400 m^2) for the accounting and collection of takyr flow, a combined infiltration system with an operational observation network, watering communication system, and laboratory and housing buildings. Precipitation water, flowing down from the takyr surface (of natural clayey catchment areas), immerses in the aeration zone via an infiltration pit to the groundwater level with salinity of 20–25 g/L, where a freshwater lens begins to form. Since the specific weight of the underground salt water is rather high, the immersing fresh rainwater with its low specific weight produces a peculiar floating lens. Experiments have indicated that at the immersion of the takyr flow from catchments with an area of 1 km^2 one can make fresh water reserves up to 10,000 m^3 which, provided they are periodically filled, ensure guaranteed water supply to a stockkeeper farm involving 10–15 people with a flock of sheep of 1,000–1,200 head.

Analysis of the hydrogeochemical data of the operational observations over a 10-year period has shown the stability of the artificial lense of freshwater in the absence of annual filling in. In the following decades, with systematic filling in, the freshwater lense could spread over the surface of the underground salt water, increasing its reserves ad infinitum. Lense freshwater can be withdrawn from wells both by a pump through a hole or by a belt water lifting device.

The prime cost of 1,000 m^3 of the local rainwater collected from takyr surface is 20 times lower than the water delivered by auto-water carriers from as far as 90–100 km. The Institute of Deserts also carried out research on the collection and spreading of rain water by means of making artificial catchments from low-cost film materials. In the short run, the Institute of Deserts is going to begin testing new, more economic, light, and heat-stable antifiltering materials that are able to form solid watertight surfaces. Such artificial catchment surfaces on any type of desert allow collecting from 1 ha for 1 year and spreading up to 700–800 m^3 of rain water compared to 300 m^3 of water from the same area of natural takyr. Theoretically, within 1 year from 1 km^2 surface of natural takyrs of artificial watertight surfaces one can collect from 5 to 30,000 m^2 of rain water.

Conclusion

Under the conditions of critical freshwater shortages, sound use of all the above mentioned alternative sources of water supply to minor water consumers in the deserts of Central Asia by means of improvement based on engineering technology remains fully urgent.

SCADA – Instrument for Increasing Water Productivity

Ismail Begimov

A key tool of integrated water resources management is the automation of water distribution systems based on the introduction of a modern system of supervisory control and data acquisition (SCADA). This system improves the quality, flexibility and reliability of water distribution management and reduces unproductive water losses.

Developing the system of automation and control of the Uchkurgan Hydroscheme was initiated in 2002 simultaneously with the launching of the IWRM-Fergana Project. The modern programmable controllers "Decont" manufactured by the company "DEP" (Russia) with home-produced water level and gate position sensors were applied in this system. One of the main tasks of the automation and control of the Uchkurgan Hydroscheme is to improve the stability of water delivery through the North Fergana Canal (NFC) and the Additional Feeding Canal (AFC) within the system of Big Fergana Canal (BFC) under fluctuating water levels in the upstream. At present, the system of automation and control of the Uchkurgan Hydroscheme does not directly receive information on flow rates at the gauging station Uchkurgan and headrace channel of the BFC; therefore a dispatcher of the Uchkurgan Hydroscheme assigns the required flow rates through the NFC and AFC depending on current flow rates and water use limits established.

The system was funded and operated by the Swiss Agency for Cooperation and Development. Specialists of the BWO "Syrdarya", SIC ICWC and SANIIRI monitored this system operation over the period of 2003–2007.

Figure 1 shows the operation mode of the automation system of the Uchkurgan Hydroscheme over the period of 2005–2006. As shown in the figure, under fluctuating water levels in the upstream, flow rates through the NFC and AFC are almost stable within the acceptable accuracy of regulation. Stability of water delivery to the NFC and AFC is ensured by the automation system of the hydro scheme at the expense of the capacity of the headrace channel and releases of excess water into

I. Begimov (✉)
Scientific Information Centre of Interstate Commission for Water Coordination,
B. 11, Karasu – 4, Tashkent, 700187, Republic of Uzbekistan
e-mail: info@icwc-aral.uz

C. Madramootoo and V.A. Dukhovny (eds.), *Water and Food Security in Central Asia*, NATO Science for Peace and Security Series C: Environmental Security,
DOI 10.1007/978-90-481-9974-7_17,

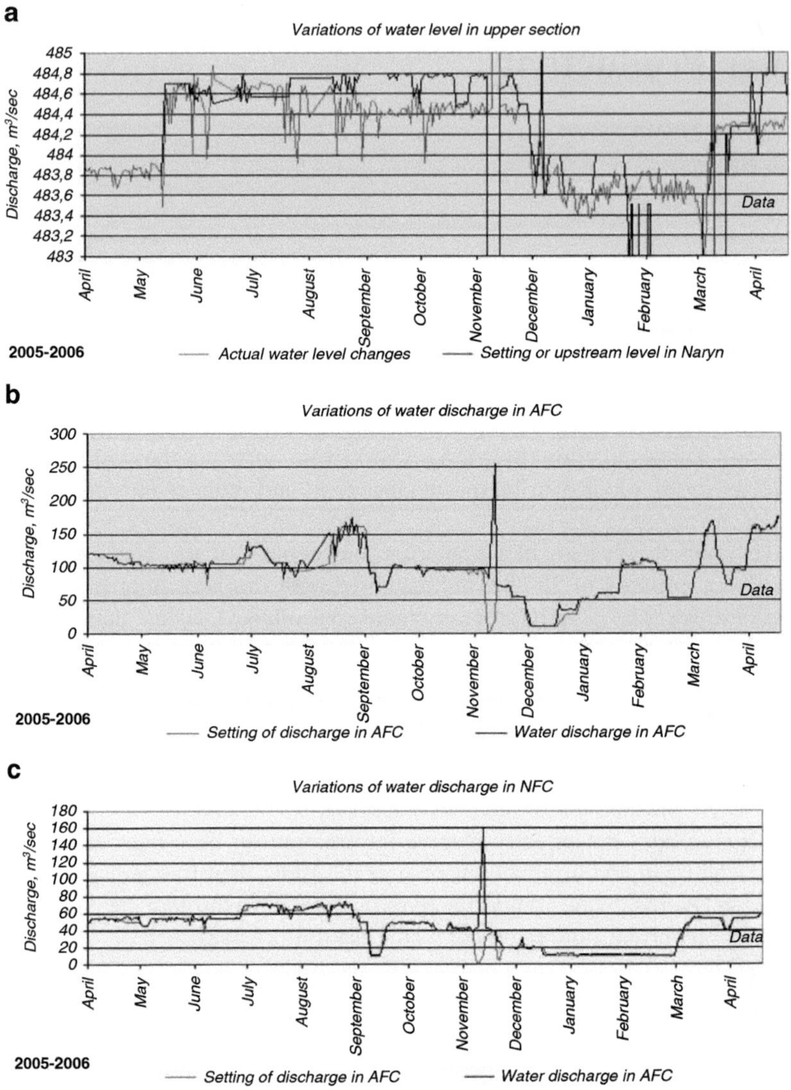

Fig. 1 Operation mode of the automation system of the Uchkurgan Hydro scheme over the period of 2005–2006

the trailrace channel. The analysis of the automation and control system operation of the Uchkurgan Hydroscheme over the whole period of operation (5 years) shows that the following qualitative and quantitative indicators of water resources management were considerably improved:

- The stability of water supply through the NFC and AFC in the Fergana Valley was improved due to the implementation of the system of automatic regulation of water levels and flow rates

- The measurement accuracy of water levels, flow rates and salinity, as well as height of opening the hydrostructure gates was raised due to the application of the modern water measurement and accounting facilities
- Dataware and the quality of water record keeping was improved through continuous collection, storage and processing data on water levels and flow rates
- The efficiency and accuracy of water management were improved through more rapid receipt and processing of data on technological progress and decision making
- Timely identification and elimination of equipment troubles in control systems and in hydrostructures was increased

An average value of deviations of actual flow rates against the established value under automatic regulating does not exceed 2% for the NFC (for the NFC 1.61%, and for the AFC 1.69%). A maximal value of instantaneous deviations of an actual flow rate against the established value for automatic regulation of a flow rate in the NFC amounts to 11.22% and 1.77% in the AFC (during the transition period).

Taking into account advantages of the automation system, the project "Automation of water distribution on pilot canals in the frame of IWRM-Fergana Project and the BWO 'Syrdarya' structures" has been proposed as further development of the IWRM-Fergana Project and its tools.

This project includes:

The basin level:

- The BWO "Syrdarya" structures

Pilot canals:

- Aravan – Akbura Canal (Kyrgyzstan)
- South Fergana Canal (Uzbekistan)
- Khoji-Bakirgan Canal (Gulyakandoz, Tajikistan)

The project objective is to put into practice the computer-aided system of regulating and operational monitoring of the water distribution process at the BWO "Syrdarya" structures and on pilot canals to ensure the supply of irrigation water to farmers in due amounts and at the proper time and to establish a system of monitoring channel inflow, flow rates and water levels at the water-balance gauging stations and water intakes.

A key task of automation and monitoring is to establish a system of management and control of canal operation, which allows:

- Improved implementation of the water use plan
- Creation of conditions for sustainable, uniform and equitable water distribution excluding unproductive water losses

Achieving these objectives will be based on the introduction of the SCADA system on the water intake and check structures, water-balance gauging stations by controlling all management objects, establishing telecommunication networks and providing computer-based data collection, processing and storage, as well as monitoring water-balance sites by observers equipped with communication facilities and vehicles.

Current Situation

Pilot canals subject to automation, have different sources of water supply:

- The South Fergana Canal is fed from Andijan Reservoir of over-year regulation
- The Akbura River, flow regulated by the Papan seasonal-storage reservoir is the water source for the Aravan- Akbura Canal
- The Khoja-Bakirgan Canal takes water from the river of the same name with unregulated flow

The existing situation in water distribution through irrigation canals and the stochastic nature of flow rates in streams impede uniform water delivery to consumers and meeting the established water use limits. Unproductive organizational water releases result from the inopportuneness and unreliability of information gathered at the gauging stations due to a lack of, or insufficient accuracy of, measuring devices for monitoring flow rates and water levels.

Automation of both the main waterworks and the system of information gathering on the water- balance gauging stations and monitoring at the water-balance sites by observers provided with communication facilities and vehicles are envisaged to ensure sustainable water distribution and providing stable and uniform satisfaction of farmers' requirements.

The System of Managing the Water Distribution Process

There are not differences of principle in the system of managing water resources on pilot canals; each republican system is represented by three levels:

1. *Basin level* where the BWO Syrdarya and the republican ministries of water resources carry out management functions. At this level, the ICWC establishes limits of water resources for the irrigation systems and controls their realization.
2. *Level of the Basin Irrigation Systems Administration, Fergana Valley Main Canals Management Organization (Uzbekistan) and Provincial Water Management Organizations (Kyrgyzstan and Tajikistan).* At this level, the plans of water use with allocating water resources per specific irrigation canals are approved taking into consideration the water use limits established and applications of farmers.
3. *Level of the Main Canal Management Organization,* at this level, irrigation water distribution over 10-day periods in accordance with an approved plan is being implemented, as well as the monitoring and adjusting of water delivery every 10 days, if necessary.

The main control station (MCS) and water-balance sites with local control stations were established on each irrigation canal in the frame of operational water distribution system. The central control station (CCS) that is the central element of water distribution management along the canal was established at the Main Canal Management Organization.

The Principle of Water Distribution Through Irrigation Canals

A key principle of water distribution through irrigation canals is planned water use based on stable and equitable meeting of consumer demand over the entire length of irrigation canals. Plans are drawn up by water resources management organizations based on applications submitted by water users and water use limits established by the Ministries. Water use plans are approved after joint reviewing by the Irrigation System Management Organizations (or Provincial Water Management Organizations), Canal Management Organizations, Canal Water Committees and representatives of water users.

Water use plans are the basis for water withdrawal plans and delivery to consumers that are being drawn up every 10 days and adjusted during the irrigation season depending on weather conditions, the general water management situation in the river basin and applications of consumers.

Extent of Automation and Controlling of Main Hydrostructures and the Monitoring System

Headworks of the pilot irrigation canals are equipped with measuring devices of the SCADA system; sensors of water level upstream and downstream of the structure and at gate positions (an extent of their lifting) are installed at all check structures. Control stations at headworks are equipped with computers and a telecommunication system that provides trouble-free communication with the central and local dispatching points and automatic data transfer according to the established mode. The following components operate automatically:

- Gates of headworks that maintain designated flow rates under fluctuating water levels in the headrace channel
- Gates of spillways operated in accordance with water levels in the headrace channel
- All information registered by sensors is illustrated at the symbolic circuits
- Protection from emergencies (self-locking of gates, exceeding a maximal levels, power cutoff, opening of power switchboard by unauthorized persons, etc.) is envisaged

The SCADA system at the main structures includes the following equipment:

- Computers
- Programmed controllers
- Input and output modules
- Level and gate position sensors
- Radio stations with antennas

Secondary canal head gates are equipped and operate similar to pilot main canal's headworks.

Fig. 2 Level sensor

Fig. 3 Gate position sensor

Controllers DECONT-182 manufactured in Russia and its input and output modules were used as programmed controllers and input/output modules. For measuring water levels, ultrasonic sensors UDU-25 (УДУ-25) with standard analogous current output (4–20 mA) were used. The temperature errors in the device are corrected with built-in thermosensor. For controlling gate positions sensors AWC 5812 with special SSI Transmitter Interface, interfacing directly with controller D-182 (DECONT) are used (Figs. 2–3).

Automation will be introduced on:

- The South Fergana Canal – ten main structures and Karkidon Reservoir's structures (72 gates and 17 control stations)
- The Aravan-Akbura Canal – three main structures (17 gates and seven control stations) and
- The Khoji-Bakirgan canal – seven key structures (43 gates and seven control stations)

Four BWO « Syrdarya » structures are also equipped with the SCADA system (46 gates and five control stations).

Water- balance gauging stations are equipped with the SCADA system (sensors of water level).

The SCADA system at water- balance gauging stations includes the following equipment:

- Programmed controllers
- Input/output modules, level sensors and radio-stations with antennas

Information on water level and flow rates is transmitted through radio communication to a local control station (LCS) of a hydro-operational site (a hydro-unit) that operates this water- balance gauging station. The following gauging stations will be subjected to automation:

- The South Fergana Canal – ten gauging stations (one at the headworks, nine at water-balance sites)
- The Aravan-Akbura Canal – four gauging stations (one at the headworks, three at water-balance sites)
- The Khodji-Bakirgan Canal – three gauging stations (one at the headworks, two at water– balance sites); and seven control stations

The Telecommunication System of the CCS and LCS

Irrigation canals are equipped with canal telecommunication systems with modern facilities for data transmitting and voice –message reports that solves the following tasks:

- Reception and transmitting telemetric information, which is formed by the automation system established in the radio-communication units of the CCS, waterworks and water-balance gauging stations
- Voice radio-communication between control stations and observers of hydro-operational sites
- Provision of the united information system of an irrigation canal based on the computerized network of transmitting, reception, processing and exchange of information between CCS and LCSs

Monitoring at Water-Balance Sites

Objects of automation and computer-aided monitoring on the pilot irrigation canals do not exceed 10% of water distribution infrastructure. Therefore, to achieve sustainable and uniform water distribution along the entire length of irrigation canals and meet user water demands, observers at water- balance sites who monitor off-takes operations play key roles.

For the purpose of efficient water resources management, the irrigation canals are subdivided into water-balance (hydro-operational) sites that are the primary level of management hierarchy. A local control station that will be equipped with computer and telecommunication means was established at each water-balance site. LCS receives information from the main structures and water-balance gauging stations and has the staff of observers who monitor water distribution at all off-takes and water diversion by pumping units. Monitoring at water-balance sites is conducted based on visual read-out of information, is transferred to the LCS by observers via their individual radiophones, and data is put into computers by hand. Off-takes at water-balance sites are divided into two-groups: "controllable off-takes" and "accountable off-takes". Off-takes (pumping units), unplanned opening or closing of which can considerably affect canal operation, refer to "controllable off-takes" and are characterized by the following parameters:

- Within the SFC system, off-takes with discharge capacity more than 100 L/s
- Within the AAC and KBC systems, off-takes with a discharge capacity more than 10 L/s

Flow rates for such off-takes can be regulated during a 10-day period; and at the same time, flow rates of off-takes with a lesser carrying capacity are not adjusted. All off-takes are "accountable ones". Water withdrawal is accounted for using water-measuring devices; however, water diversion through small off-takes with a discharge capacity less than 5 L/s and watermills is accounted for according to their rated discharge capacity.

Water withdrawal by pumping units is calculated taking into account the number of pumping units (PU) under operation and their nameplate capacity and audited according to read outs of an energy meter (Table 1).

The number of daily observations is established depending on the duration of daylight hours: during the growing season – four times a day and during the off-vegetation period- three times a day. Time spent by an observer at one structure was estimated based on the duration of each elementary procedure:

At off-takes:

1. Readout of an indication of a water-level staff installed in the headrace channel
2. Readout an indication of a water meter staff and determination of a flow rate using the design chart
3. Data transmitted to a dispatcher and
4. Data recorded into the field book

At the pumping units:

1. Visual definition of the number of pumps under operation
2. Reading indication of an energy meter
3. Data transmitted to a dispatcher and
4. Data recorded into the field book

Table 1 Structures under monitoring

Irrigation canal	Number of off-takes		Total water withdrawal		Small pumping units and watermills (Q < 5 L/s)		
	Total	Including PU	Q (m^3/s)	% of Q_{st}	Unit	Q (m^3/s)	% of Q_{st}
SFG	162	67	92	92	68[a]	3.89	2.95
AAC	62	5	28.8	87	108	0.54	2
KBC	46	4	32.6	80.2	14	0.07	0.2

[a] Off-takes with a discharge capacity less than 100 L/s were included into the group of off-takes non-controllable during a 10-day period within the SFC system.

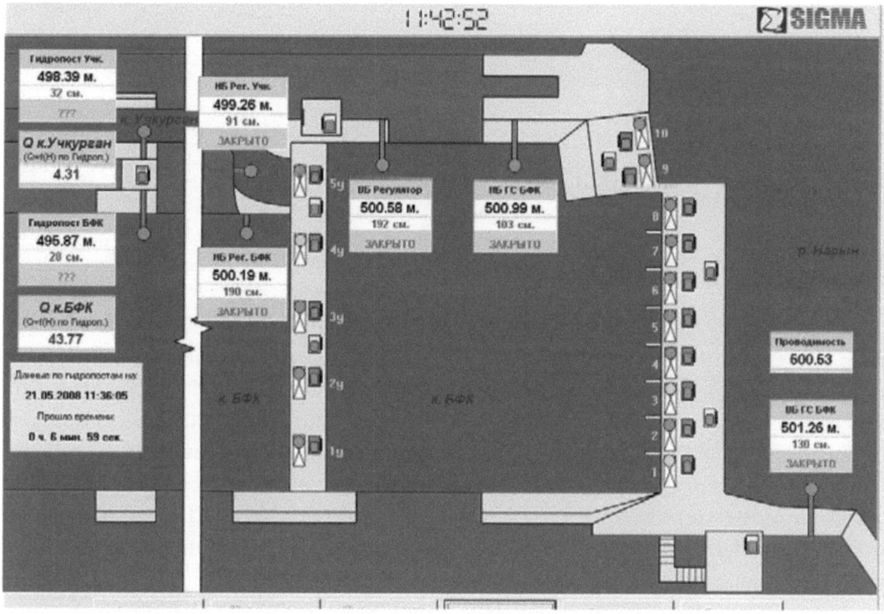

Fig. 4 Main symbolic circuit of the system of automation and controlling of the BFC headworks

Observers are provided with vehicles (mopeds) and radio-telephones. The number of observers was specified on the basis of a length of water-balance sites, number of off-takes, and normative working hours. Project works were monitored and coordinated by SIC ICWC and SANIIRI. The system of automation and controlling of the Naryn-Karadarya Waterworks Management Organization of BWO "Syrdarya" was put in pilot operation in July 11, 2006.

A usable automation system interface was developed, where all automation functions were represented in the symbolic circuits of the system. Figures 4 and 5 show the main symbolic circuits of the system of automation and controlling of the headworks of the BFC and the system of transferring data on BWO "Syrdarya" structures respectively; based on which dispatchers of LCS and CCS can estimate the actuality of presented data.

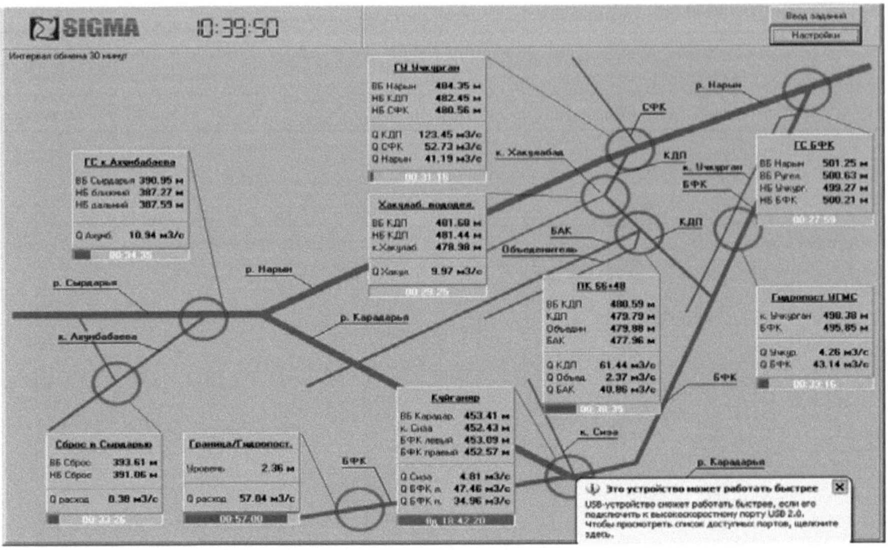

Fig. 5 Symbolic circuit of data transfer system

It is necessary to note that at the head of the BFC and site of the AFC for the first time downstream regulation system has been implemented. The system enables interlinked automatic operation of the headworks of the BFC with check structures on DP 3 + 75 on BFC and headworks of AFC canal with check structures of the AFC on DP15, DP 66. Downstream regulation systems automatically maintain established water discharges in the side outlets and downstream water level in control structures.

Existing transfer channel, simplex radio stations ICOM FC F-410, is used for transmitting data between CCS and the Uchkurgan Hydroscheme; it was earlier used in the automation and controlling project of the Uchkurgan Hydroscheme to exchange information. Continuous data from CS objects of the Uchkurgan division are transmitted at least once every 10 min through the GPRS network. Moreover, time passed since the last successful data transfer is also displayed on the computer allowing the dispatcher to estimate the actuality of submitted data.

In May 2008 the system of automating and controlling the Naryn-Karadarya Waterworks Management Organization of the BWO « SYRDARYA » was put into industrial operation. The system of automating and monitoring the AAC was put into pilot operation in October 2008 and in October 2009 for the KBC and SFC.

The Cost Effectiveness of « Fergana Valley Canal Automation » Project

Indicators of the cost effectiveness of "Fergana Valley Canal Automation" project that resulted from analyzing project parameters are the following:

Table 2 Evaluation of cost effectiveness of "Fergana valley canal automation" project

	Operating costs (000. USD)		Investments (000. USD)		Net profit		Economic indicators		
Objects	Before	After	SDC	MAWR	Water volumes (mln. $m^{3)}$	Cost (000. USD)	Cost recovery (years)	NPV over 15 years	IRR (%)
BWO "Syrdarya"	74	87.36	305	40	38.8	232.8	2	1,039	59
SFC	131	155	725	117	63	378	7	1,224	36
AAC	59	64	235	30	7	69.8	9	121	19
KBC	68	71	280	75	6.47	38.8	7	93	24
In total	332	377.36	1545	262	115.27	719.4	6	2,477	

- Investments – USD 1,545,000 funded by the SDC and USD 262,000 budgeted by the water management organizations of the republics
- Operating costs – USD 332,000 and 377,360 prior to and after project implementation respectively
- Annual net profit due to water savings amounts to USD 719,000 (115.27 million m^3 at a water price of 0.006 USD/m^3)
- The cost recovery- 6 years
- The net present value- USD 2,477,000 and IRR = 32%. Results of the economic analysis are given in Table 2

Based on results of the economic analysis of existing systems of operation it can be noted that the introduction of the system of automation and monitoring of the water distribution process within the irrigation systems in Central Asian countries is the cheapest measure for water resources saving compared with other technical solutions such as canal lining or other measures preventing water seepage losses.

The small-scale enterprise "SIGMA" (Kyrgyzstan), whose production is comparatively cheap, simple to operate and accessible for procurement, taking into consideration available operational and servicing personnel in our region, was selected as the leading Contractor for constructing the system of automation and monitoring. Installation and construction works on preparation of hydrostructures and control stations were accomplished by the efforts of Naryn-Karadarya Waterworks Management Organization of the BWO « Syrdarya » and water management institutions of Central Asian countries.

Conclusions

As a result of the implementation of automation and monitoring systems at BWO « Syrdarya » structures and pilot canals the following were achieved:

- Improvement of accuracy in measuring water level, discharge and salinity, as well as in opening gates of hydrostructures through application of modern water measurement and accounting facilities (discharge measurement error decreased from 5–10% to 2–3%)

- Improvement of information provision through continuous collection, storage and processing of measured values of water level and discharge in computers
- Increase of operational efficiency and accuracy of water management through more rapid receipt and data processing on technological processes and decision making
- Stabilization of the delivery of fixed discharge through the main canals of Fergana Valley
- Timely identification and elimination of equipment troubles in control systems and in hydrostructures

It is necessary to note that the systems of automation and control installed at BWO "Syrdarya" structures and pilot canals AAC, SFC, and KBC have improved operations, making the jobs of operating staff easier, and have improved water allocation in the large canals such as AFC, NFC, Big Andijan Canal, Khakulabad Canal, Akhunbabaev Canal, AAC, SFC, and KBC. Conditions for real system monitoring by the BWO and pilot canals, its territorial bodies and for receiving reliable information on water resources by all stakeholders were created.

At present, SIC ICWC together with SANIIRI, BWO « Syrdarya » and « Amudarya » are preparing proposals on further implementation of analogous systems in the rest of BWO « Syrdarya » structures, development of automation and monitoring systems in the small rivers of the Fergana Valley and working out a feasibility study on the development of analogous systems for BWO « Amudarya » structures.

Central Asia Regional Water Information Base – CAREWIB

Iskander Ferdinandovich Beglov

Water resources management in transboundary river basins involves multifaceted interactions among various stakeholders at all levels of the water management hierarchy and requires a shared vision of sustainable water resources management and development in Central Asia.

Since 1991 the newly formed independent states of the Central Asian region have been faced with an acute necessity to develop regionally. Access to and exchange of information are the key instruments to maintain effective and efficient regional cooperation, but the lack of information about water/land resources use and the socio-economic situation in the Aral Sea Basin has been an obstacle for making adequate short- and long-term decisions on transboundary water resources management and on implementation of relevant policy.

Recognizing the need for cooperation on problems such as water allocation in the Aral Sea Basin resulted in the signing of an interstate agreement on establishing the Interstate Commission for Water Coordination (ICWC) in charge of regional water resource management within the Aral Sea basin (1992). The Scientific-Information Centre (SIC ICWC) was founded as an ICWC executive body to coordinate regional water management cooperation and improve information exchange among the member states.

In this context, the development of a regional information exchange system became one of the most important components of the efforts to improve water resources and ecosystem management at the regional, national and province levels. ICWC has given considerable attention to the use of many tools available in the region, which facilitate information exchange among various water stakeholders and are suitable for different settings and types of people.

The application of all IWRM tools available by 2003 showed that more advanced technologies and sophisticated methods of communication between water stakeholders were needed – such as a special interactive web-based *Basin Information*

I.F. Beglov (✉)
Scientific Information Centre of Interstate Commission for Water Coordination,
B. 11, Karasu – 4, Tashkent, 700187, Republic of Uzbekistan
e-mail: info@icwc-aral.uz

C. Madramootoo and V.A. Dukhovny (eds.), *Water and Food Security in Central Asia*, NATO Science for Peace and Security Series C: Environmental Security,
DOI 10.1007/978-90-481-9974-7_18,

System. Such a system with constantly updated information can be used to identify the best management options for specific conditions, tasks, planning zones in line with overall river basin management goals and targets. It was also acknowledged expedient to introduce an interactive Geographical Information System (GIS) designed for use within agencies or for targeted partners in a water management context. *SDC – being an active promoter and initiator of implementing IWRM principles in the Aral Sea basin – decided to support* SIC *ICWC to develop work along these lines in collaboration with the UNECE and UNEP/GRID-Arendal Office in Geneva. As a result the Central Asia Regional Water Information Base (CAREWIB) Project was launched on December 2003.*

Despite the complexity of the current socio-economic situation in Central Asian countries, water resource use and management at both on-farm and irrigation system/basin levels are undergoing positive changes, quickly adapting to current developments, and meeting complex challenges of the current transition to a market economy. By enhancing information exchange in the water sector, the CAREWIB Project has undoubtedly contributed to this favorable development. Based on the good will and mutual understanding on the part of all ICWC members, CAWater-Info Portal and Information System (IS) were created during Phase-1 resulting in unique information products that have no analogues in Central Asia.

The CAREWIB Project is implemented in support of the ASBP-1 approved by the decision of the Heads of State on 11 January 1994 (Project 2 – "Database and management information system for water and environment") and ASBP-2 approved by the IFAS Board on 28 August 2003 (Item 6 – "Reinforcement of material/technical and legal basis in interstate organizations, development of the regional information system designed to manage water resources of the Aral Sea basin"). The CAREWIB Project also supports the Orhus Convention implementation.

During the transition period (1 January – 31 July 2007) the project's activity assessment was implemented by independent experts; the outcomes of Phase I and expected results of Phase II were discussed with the main concerned users; and the project document was finished on the basis of the obtained recommendations. Phase II was started 1 August 2007and its finishing date is 30 June 2010.

CAWater-Info web portal

Structurally, the Portal consists of a "central" CAWater-Info website (*www.cawater-info.net*), a group of official websites, a group of project websites, on-line database and knowledge base and the section "Water World". In addition, traditional services such as daily updated news, calendars of events, forums and so on are available on the portal.

Central Website of CAWater-Info Portal

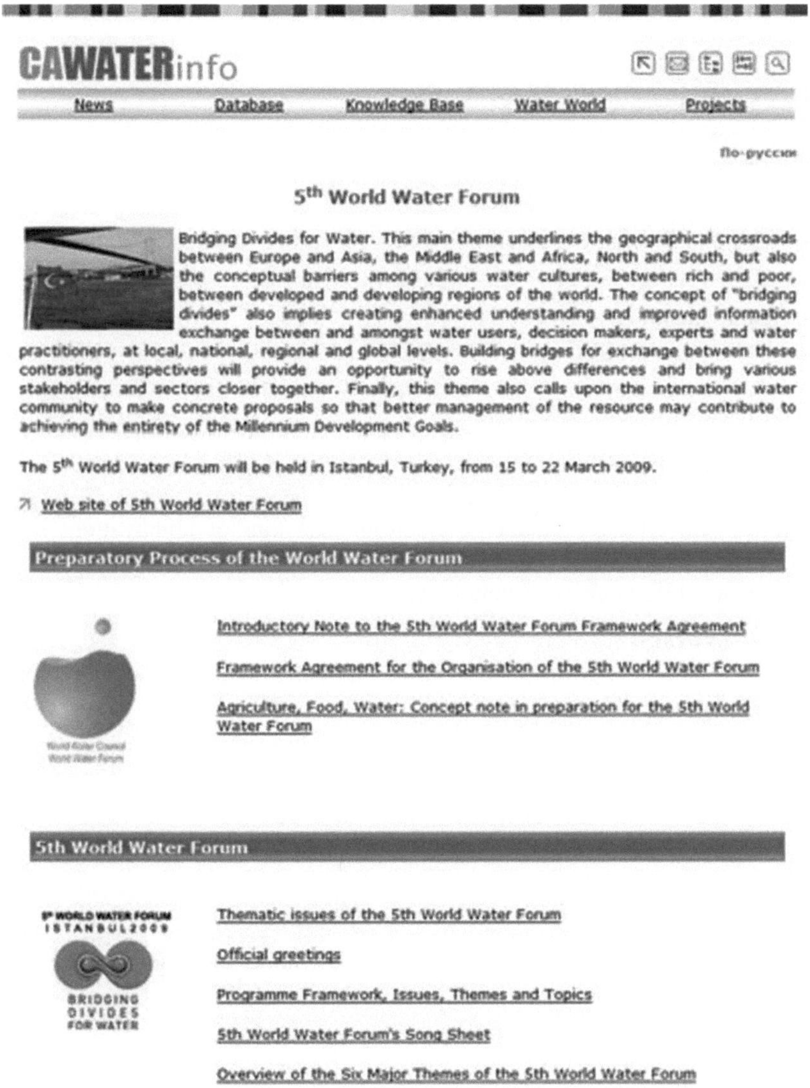

CAWATERinfo

News Database Knowledge Base Water World Projects

По-русски

5th World Water Forum

Bridging Divides for Water. This main theme underlines the geographical crossroads between Europe and Asia, the Middle East and Africa, North and South, but also the conceptual barriers among various water cultures, between rich and poor, between developed and developing regions of the world. The concept of "bridging divides" also implies creating enhanced understanding and improved information exchange between and amongst water users, decision makers, experts and water practitioners, at local, national, regional and global levels. Building bridges for exchange between these contrasting perspectives will provide an opportunity to rise above differences and bring various stakeholders and sectors closer together. Finally, this theme also calls upon the international water community to make concrete proposals so that better management of the resource may contribute to achieving the entirety of the Millennium Development Goals.

The 5th World Water Forum will be held in Istanbul, Turkey, from 15 to 22 March 2009.

Web site of 5th World Water Forum

Preparatory Process of the World Water Forum

Introductory Note to the 5th World Water Forum Framework Agreement

Framework Agreement for the Organisation of the 5th World Water Forum

Agriculture, Food, Water: Concept note in preparation for the 5th World Water Forum

5th World Water Forum

Thematic issues of the 5th World Water Forum

Official greetings

Programme Framework, Issues, Themes and Topics

5th World Water Forum's Song Sheet

Overview of the Six Major Themes of the 5th World Water Forum

News of Central Asia

Digest of daily updated information materials; covers the following subjects: water resources, environment, energy, socio-economic development, etc.
www.cawater-info.net/news/

CAWater Info|News
An official bulletin of the CAWater-Info portal published under the 'Central Asia Regional.Water Information Base (CAREWIB)' Project.
www.cawater-info.net/news/carewib-news.htm

Reviews and analytical articles
Reviews and analytical articles on the portal subjects.
www.cawater-info.net/review/

News: Afghanistan
News section covering the situation of water resource use in Afghanistan.
www.cawater-info.net/news/afghanistan/

Facts about water
In the section, materials from the UNESCO Water Portal relating to water resources use in different human activity areas translated into Russian are available.
www.cawater-info.net/news/water/

New technologies
The section of the portal that covers new developments in water resources and agriculture; includes an archive of electronic versions of "Innovations in water management" bulletin.
www.cawater-info.net/news/technology/

Calendar of events
Calendar of seminars, symposiums, conferences, etc.; includes plan of operation of the ICWC Training Center.
www.cawater-info.net/events/

Catalogue of water and environmental sites
A set of sites on water and environmental subjects carefully selected by hand and sorted out by subjects.
www.cawater-info.net/sites/

Calls
Announcements about calls, in which scientists from Central Asia can take part.
www.cawater-info.net/calls/

Database

Includes useful information: addresses of water and environmental organizations in Central Asia (including NGOs), brief information on the Aral Sea basin states, a set of links to external (outside) databases and other useful information.
www.cawater-info.net/bd/

On-line data of the BWO Amudarya

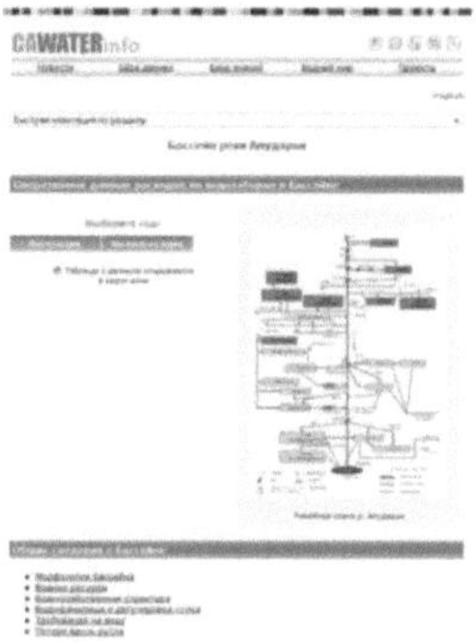

On-line data of the BWO Syrdarya

Database of the Aral Sea

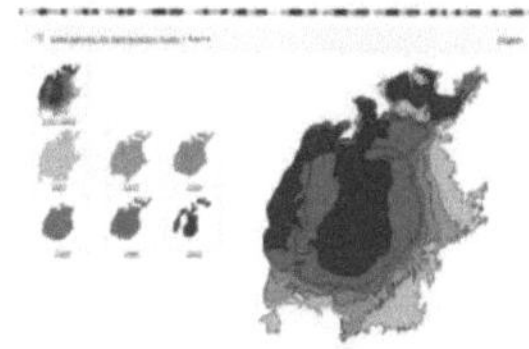

Analysis of water-related situation in river basins of Amudarya and Syrdarya

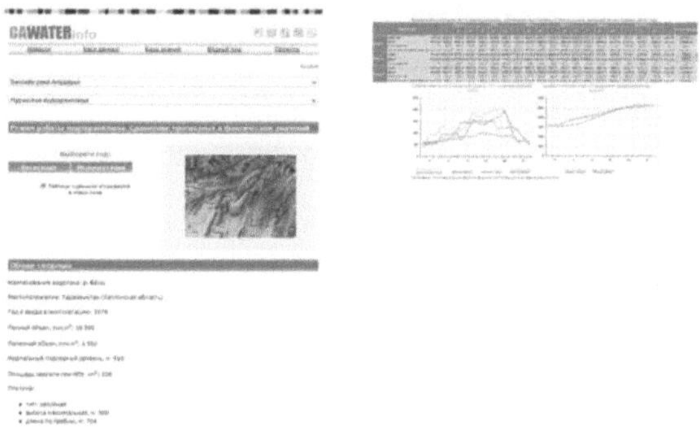

Regional Information System
Description of the CAREWIB IS
www.cawater-info.net/careweis/

On-line data on discharges per water intake in the Amudarya River basin
On-line database on actual 10-day water withdrawals in the Amudarya River basin. The section also contains general information on the basin: basin morphology; water resources; water management structure; reservoirs and flow regulation; water requirements; losses along the channel.
www.cawater-info.net/amudarya/

On-line data on discharges per water intake in the Syrdarya River basin
On-line database on actual 10-day water withdrawal in the Syrdarya River basin. The section also contains general information on the basin: basin morphology; water resources; water management structure.
www.cawater-info.net/syrdarya/

Analytical Reviews
Analytical reviews allow implementing the integrated assessment of water management situation in the river basins of Amudarya and Syrdarya, and their sites including:

- The projective (look-ahead) and actual parameters of tributaries, releases, reservoirs' volume, water intakes
- Deviation between projective (look-ahead) and actual parameters with water availability for water intakes
  www.cawater-info.net/analysis/water/

The Aral Sea Basin
General information: Location, geomorphology, landscape, climate; Water resources; Formation of the surface flow; Surface water resources quality; Lakes and depressions; Groundwater: reserves and use; Waste and drainage water; Dams and hydropower; Lands; Irrigated lands; Salinization and drainage; The Aral Sea basin in figures. History of the Aral Sea: from ancient times to the present.
www.cawater-info.net/aral/

Database on the Aral Sea
Map; precipitation; level; wind speed; water temperature; water mass volume; volume of water from river; relative humidity; evaporation from water surface.
www.cawater-info.net/aral/data/

Water resources of Afghanistan
Surface water; hydrological study; surface water resources; hydrogeology and groundwater
www.cawater-info.net/afghanistan/

Database on rivers of Afghanistan
Data on 16 rivers in Afghanistan gathered by the State hydrological; institute (Russia)
www.cawater-info.net/afghanistan/data/

The Aral Sea Basin Management Model
Description of the Aral Sea Basin Management Model (ASB-MM)
www.cawater-info.net/asbmm/

UNECE Environmental Performance Review
UNECE Conventions; Environmental performance reviews; Special programs.
www.cawater-info.net/unece/

Reports on the state of the environment of GRID-Arendal
Reports on the state of the environment of UNEP/GRID-Arendal
www.cawater-info.net/grida/

Knowledge Base

Knowledge Base is the organized body of knowledge pertaining to some object domain. Knowledge is the result of reality cognition tested by practice. In other words, knowledge is truth, facts, principles and other knowledge objects accumulated by humankind. Therefore the cognizable information of documents, books, articles, reports are in the knowledge base (as opposed to the database).

Contains processed knowledge structured by sections: Electronic Library of SIC ICWC; The publications of the month; Central Asian newspapers and magazines; Photolibrary; Thematic knowledge bases; Glossary; Bibliographic database.
www.cawater-info.net/bk/

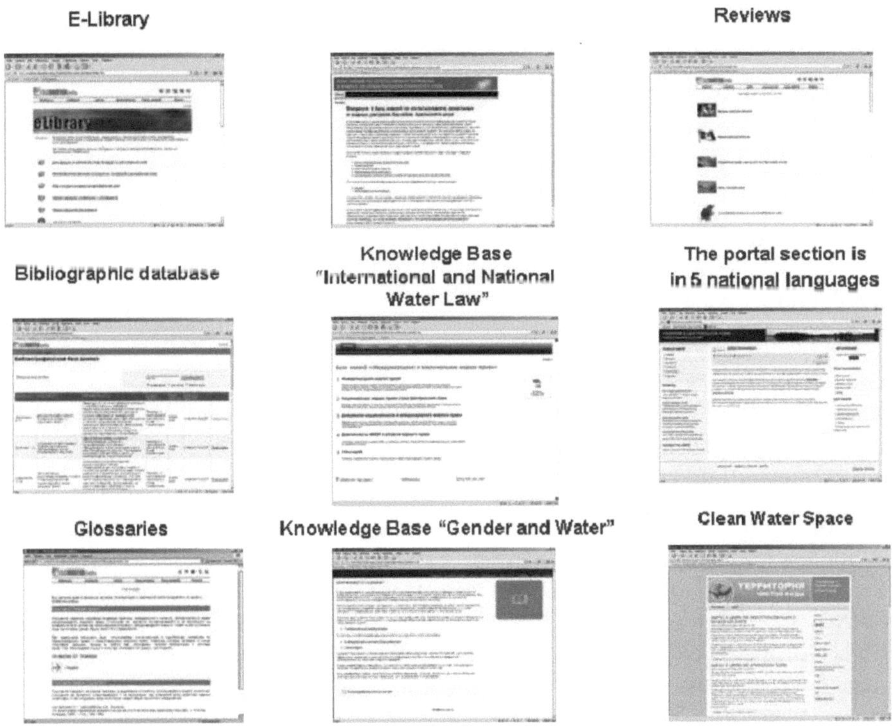

Electronic Library

Information structured by sections: Declarations and Statements of the Heads of Central Asian States; Intergovernmental Agreements of Central Asian States; Constitutions of Central Asian States; International Conventions and Agreements; International Declarations; ICID Declarations; ICWC Agreements; International Water Law; National Laws on Water; National Laws on Land; Environmental Laws of the States of Central Asia and Caucasus; Aral and Priaralie; ICWC Bulletins; Reviews of SIC ICWC; Information Digests of SIC ICWC; Collected Articles on Legal Issues of SIC ICWC; Publications of the ICWC Training Center, Publications of GWP CACENA; Publications of ICC; Books; Articles/Reports; Abstracts; Leaflets; Maps; World Water Forums; CAREWIB Project's publications, Publications of EC IFAS; Publications of the IWRM-Fergana Project; Posters; Use of Library free of charge.
www.cawater-info.net/library/

The publication of the month

Information on released publications.

www.cawater-info.net/publications/

Central Asian newspapers and magazines
Catalogue of newspapers and magazines being issued by water agencies and organizations in Central Asia.
www.cawater-info.net/publications/magazines/

Photolibrary
Water objects of Central Asia.
www.cawater-info.net/photolibrary/

Knowledge Base "International and National Water Law"
The knowledge base contains generalized information on international water law and water law of the Central Asia States.
www.cawater-info.net/bk/ water_law/

Knowledge Base "Impeller pump supply regulation"
The knowledge base contains generalized information on impeller pump supply regulation.
www.cawater-info.net/bk/pumps/

Glossaries
The Glossaries contain the main legal terms, definitions and notions used in international water law and water management.
www.cawater-info.net/bk/glossary/

Bibliographic Database
While creating bibliographic database "Land and water resources use in the Aral Sea basin", the long-term experience of the Scientific-Technical Library of SANIIRI and Information-Publishing Division of SIC ICWC, as well as of the library of the ICWC Training Center being formed has been used; contains more than 2,400 records.
www.cawater-info.net/biblio/

Websites of Projects

Donors of the Aral Sea basin
A site dedicated to donors of the Aral Sea basin (in English)
www.cawater-info.net/donors/

Website of the RIVERTWIN project
RIVERTWIN project: "A Regional Model for Integrated Water Management in Twinned River Basins"

The project objective is to develop an integrated regional model for strategic planning of water resources management (WRM) in twinned river. Three river basins were selected for modeling in different regions (the Neckar river, the Oueme river, and the Chirchik river in Germany, Benin, and Central Asia, respectively).

Based on modeling and subject to social, economic, and ecological sustainability indicators, it is planned to elaborate integrated development scenarios for each of the river basins as a basis for basin water management plans.
www.cawater-info.net/rivertwin/

Website of the TWINBASIN^{XN} project

TWINBASIN^{XN} project: "Promoting Twinning of River Basins for Developing Integrated Water Resources Management Practices".

The main objective is to promote research and development in the integrated water management sphere by encouragement of tightening ties among basin organisations.
www.cawater-info.net/twinbasinxn/

Website of the UCC-Water project

UCC-Water project: "Speedup of IWRM-2005 Objectives Implementation in Central Asia". Sub-regional Program for the Central Asia within the Framework of UNEP Support for achieving the Johannesburg Plan of Implementation target – "Integrated Water Resources Management and Efficiency Plans by 2005". The UNEP is designed for sub-regions. However, as a part of the sub-regional working program, assistance for development of the national road maps and the IWRM working programs by 2005, is provided to the selected countries (the Kyrgyz Republic, the Republic of Tajikistan, and the Republic of Uzbekistan).
www.cawater-info.net/ucc-water/

Website of the ADB RETA 6163 project

Technical Assistance « Improvement of Shared Water Resources Management in Central Asia » (ADB RETA 6163) is financing by Asian Development Bank. The overall goal of the TA is to achieve sustainable and efficient water management in the region. The purpose of the TA is to help the CARs strengthen their cooperation in the management of shared water resources in the Aral Sea Basin and other transboundary basins.
www.cawater-info.net/reta/

Website of the Information-Consulting Center for the European Commission's Seventh Framework Program in Central Asia

The main activity of the Information-Consulting Center (ICC) for the European Commission's Seventh Framework Program (7FP) in Central Asia is to overcome isolation of Central Asian researchers from European science and involvement them into the scientific and technological cooperation with European countries within context of the 7FP objectives. Announcements about 7FP calls and information helping to prepare and submit proposals are placed on the website.
www.fp7.cawater-info.net/

Website of the Network "GWANET – Gender and Water in Central Asia"

Materials of the ADB project "GWANET – GENDER AND WATER IN CENTRAL ASIA" are placed on the site. The objective of the GWANET is to promote

the gender equality within water resources management in Central Asia by strengthening the women role in the decision-making process.
www.gender.cawater-info.net

Website of the "Dialogue on Water and Climate" project
Research materials of climate impact on water resources in the Aral Sea Basin are placed on the web site.
http://dialogue.icwc-aral.uz/first.htm

Website of the "Integrated Water Resources Management in the Ferghana Valley" project
The web site is devoted to propaganda of the integrated water resources management in Central Asia by means of the example of the most important project implementing in the region – "IWRM-Ferghana".
http://iwrm.icwc-aral.uz/

Water World

Towards the fifth World Water Forum

About the progress in the preparation of Central Asian states to the fifth World Water Forum
www.cawater-info.net/5wwf/
Activities of international organizations
This section is dedicated to international non-governmental organizations with a view to broader involvement of Russian-speaking audience in their activities.

www.cawater-info.net/int_org/

World Water Council

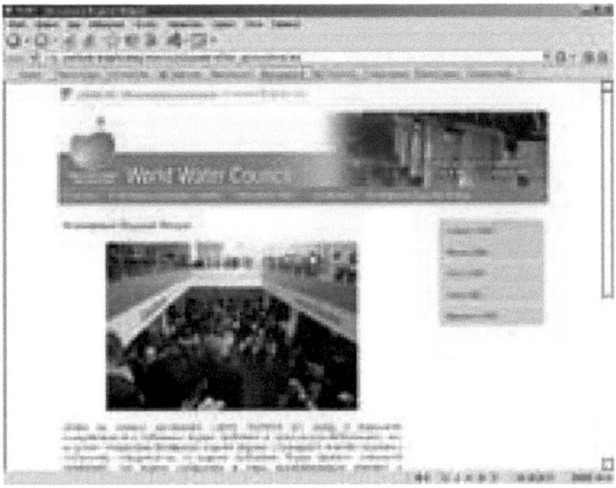

About the activities of the World Water Council. SIC ICWC also supports the section in Russian on the WWC web site.
www.cawater-info.net/int_org/wwc/
www.worldwatercouncil.org/index.php?id = 1&L = 3

Asia-Pacific Water Forum

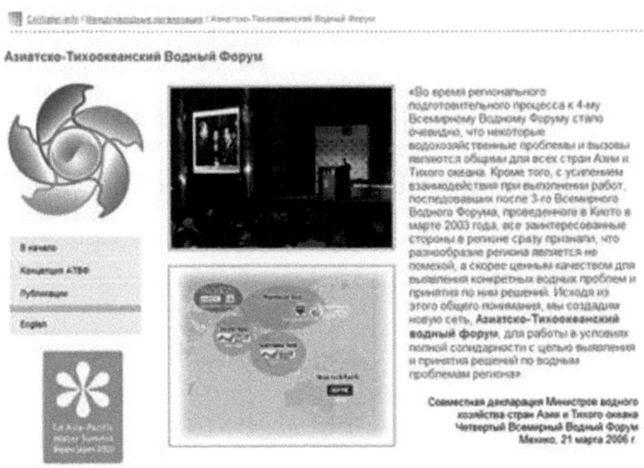

About the activities of the Asia-Pacific Water Forum
www.cawater-info.net/int_org/apwf/

International Network of Basin Organizations

About the activities of the International Network of Basin Organizations. SIC ICWC also supports the section in Russian on the INBO web site.
www.cawater-info.net/int_org/inbo/
www.inbo-news.org/ru/

International Commission on Irrigation and Drainage

About the activities of International Commission on Irrigation and Drainage
www.cawater-info.net/int_org/icid/

Official Websites

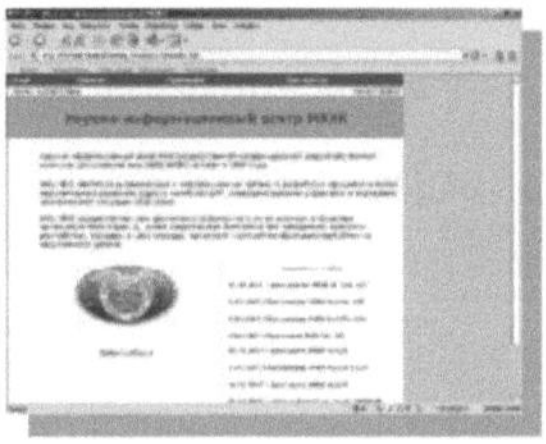

Website of the Interstate Commission for Water Coordination in Central Asia
Contains information on the history of establishment and day-to-day activities of ICWC, brief information on all meetings, reports on the largest conferences with participation of ICWC, reference data (addresses, telephone numbers) of all ICWC members. Home pages of ICWC executive bodies such as BWO "Amudarya", BWO "Syrdarya" and CMC ICWC are available at the site.
www.icwc-aral.uz

Website of the Scientific-Information Center ICWC
Contains information on SIC ICWC: regional activities, international cooperation, including activities of international organizations, projects, publications, reference data (addresses, telephone numbers) of SIC and its branches in Central Asian states.
http://sic.icwc-aral.uz

Press Releases of ICWC
On-line archive of Press Releases of ICWC published since 2002.
http://sic.icwc-aral.uz/releases/

Website of the ICWC Training Center
Contains information on activities of the ICWC Training Center, including subjects of lectures, minutes of discussions and decisions.
http://tc.icwc-aral.uz/

The overall scheme of information exchange and generation shows that in addition to wide current and basic information exchange with correspondent and user cells in the five Central Asian states, a considerable part of the informational flow is formed owing to receiving materials from various organizations and projects as well as translations and exchange of documents obtained from foreign and international organizations – constant partners of SIC ICWC.

General statistics on the CAWater-Info domain visits in 2008

Month	Unique visitors	Number of visits	Pages	Hits	Bandwidth
Jan 2008	10,216	13,386	35,019	137,796	1.80 GB
Feb 2008	9,843	13,100	33,345	133,530	1.88 GB
Mar 2008	10,482	13,535	36,655	142,272	2.36 GB
Apr 2008	11,850	15,982	47,650	161,779	2.41 GB
May 2008	11,097	15,017	51,277	166,517	2.89 GB
Jun 2008	10,349	13,902	45,629	149,512	2.46 GB
Jul 2008	8,119	11,545	49,622	155,356	2.64 GB
Aug 2008	6,894	10,722	42,281	134,645	2.00 GB
Sep 2008	9,953	14,805	54,944	186,948	2.33 GB
Oct 2008	13,988	19,582	62,996	228,252	2.80 GB
Nov 2008	15,488	21,645	81,814	260,925	3.93 GB
Dec 2008	16,921	24,173	84,213	274,236	3.67 GB
Total	135,200	187,394	625,445	2,131,768	31.18 GB

About 10,000–15,000 people visit the portal monthly (and also 2,000–3,000 people visit the domain – www.icwc-aral.uz):

Geography of visits:

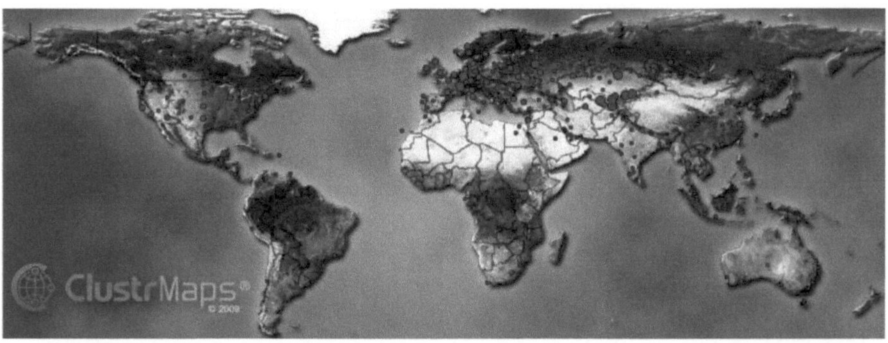

In parallel to the maintenance of the Web Portal, the project staff produces a number of periodical publications, reports and monographs and widely disseminates them in both electronic and printed forms.

Information System on Water and Land Resources

At the end of the Project's Phase I, the regional information support system became fully functional and can at present provide decision-makers, stakeholders and a broader public with timely, regular and relevant information. The information includes most water management issues, water resources and other water-related problems such as hydropower and environment, as well as measures undertaken to achieve sustainable water resources management.

The Information System developed by the project personnel in collaboration with BWO and water management organizations in the basin is a set of software and hardware linked to a database that enables information to be searched, received, stored, protected, processed and transmitted using specially developed methods. This system is a practical tool for integrated assessment of the water situation: available water resources and their allocation among river reaches, provinces and water-management systems; operation modes of reservoirs and hydro power stations; losses, shortage, imbalance; environmental releases; water quality indices, etc.

In combination with the Aral Sea Basin Management Models (ASB-mm) the system will in the future enable the regional and national organizations to use a common "information language" that will raise the reliability of used data and consequently the efficiency of water resources management.

The Information System enables users to obtain, process and analyze necessary information in the blocks: "Land", "Water", and "Economics". The existing DB is an effective system designed to provide users with reliable water-related information (e.g. *land resources, hydropower, and economy*). During the IS filling by project correspondents from the riparian countries of the basin, data on 90 parameters for the 1980–2007 period were collected. The Information System includes more than 1,200 parameters as a whole.

Currently the new blocks – "Water Supply", "Return Water", "Rehabilitation measures" and "Energy" – are being placed into the Database.

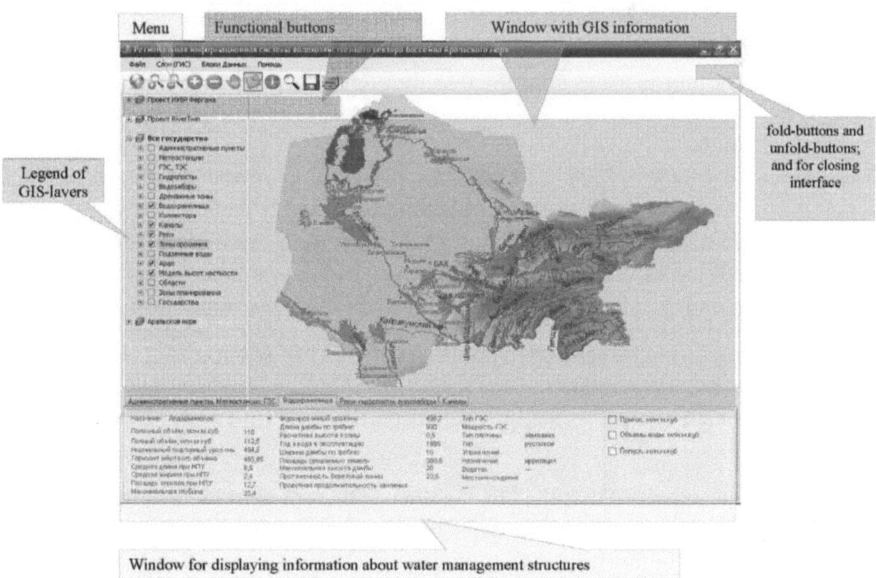

The project personnel have developed GIS maps for each of CAR. The GIS block is directly connected with the DB making it possible to visually identify a specific object (river, canal, water intake, gauging station, hydropower plant, heat power plant, etc.), and look through information related to a selected object on the map. The foundation has been laid for future work with application of GIS methods: a digital elevation model has been elaborated for Uzbekistan, South Kazakhstan, Kyrgyzstan, Tajikistan and Turkmenistan.

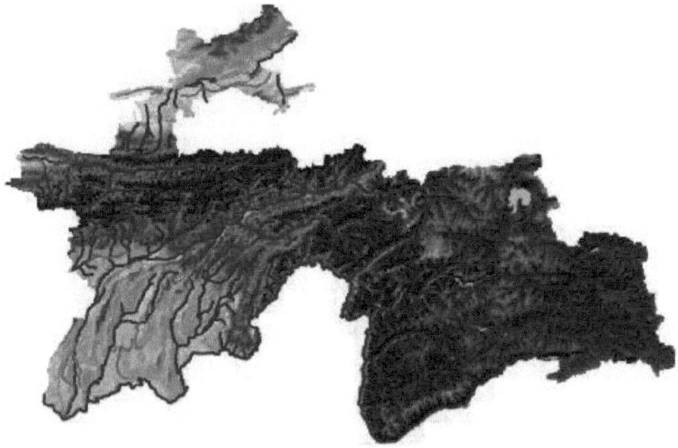

Digital map of Tajikistan

Though the Information System is presently primarily a tool for management and cooperation within the framework of ICWC, part of the information from the Information System is accessible for free through CAWater-Info Portal.

Analytical Tools

The analytical block, which was used for preparation of the monthly analytical reports for ICWC members and for publishing reviews for the general public on the portal was created in the regional information system Analytical reviews allow the implementation of the integrated assessment of water management situations in the river basins of Amudarya and Syrdarya, and their sites including:

- The projected (look-ahead) and actual parameters of tributaries, releases, reservoir volumes, water intakes
- Deviation between projected and actual parameters with water availability for water intakes

In future it is expected that the DB together with a set of models being developed and already working will enable each of the participants – water management, planning and other authorities in the countries, BWOs – to predict their own options for future development and regimes of water releases and allocation in the current context, in order to assess the impact of their own actions on other countries and separate planning zones.

At the same time, there is an opportunity to permanently assess water use efficiency for all participants in joint management and specify volumes of non-productive water withdrawal.

Regular information and forecasting on the status of natural water resources, comparison of day-to-day forecasts and actual data on water resource use and channel balances for major rivers in the region will make it possible to improve the quality of their management, and generate mutual trust, a spirit of solidarity and joint responsibility between countries and sectors of the economy.

Acknowledgements: I express my thankfulness to Prof. Dukhovny V.A. for his valuable comments during the preparation of this report, and also to Mr. Sorokin D.A. and Mr. Turdybaev B.K. for their assistance.

Steady Management of Transboundary Drainage Water

Myratgheldy Akmammedov

The formation of huge amounts of drainage water is characteristic of irrigated agriculture in the Central Asian region. During the intensive development of new irrigated lands, drainage water amounted to 39–40 km^3out of a total surface water resource of 110–115 km^3. In recent years, a decrease in drainage flow volume to 32–34 km^3 has been observed due to forced reduction of specific water supply per irrigated hectare because of excessive expansion of irrigated areas, development of industry and other water-using sectors of the economy, and population increases.

Water and land use dynamics in the Aral Sea basin (Table 1) show that from 1960 to 2006 the population of the region increased more than three times to about 45 million people and the irrigated area increased about twice to almost 8.5 million ha. The total water intake amounts to about 166 km^3, including 96 km^3 for irrigation. The specific water intake per hectare of irrigated land has declined continuously and was 11,350 m^3 in 2006.

Of the total volume of collector and drainage water (CDW) (Table 2), around 51% (18–20 km^3) is discharged to rivers and carries over 110–120 million ton of salts. Over 36% of the CDW, about 16–17 km^3, is discharged to natural depressions and is lost to evaporation and filtration and only a minor part (13 %, or 4–5 km^3) is reused for irrigation.

In the overall water drainage of return water, industrial and household waste waters are, as a rule, drained together, especially when they are discharged to river basins or local natural depressions. Therefore, when developing measures for the utilization of return waters, it is reasonable to consider the two together and keep in mind that the waste water of an industrial plant is more polluted.

CDW, a by-product of irrigated agriculture, can serve as an additional water reserve to be used for the irrigation of salt-resistant agricultural crops; as a water body for the maintenance of biological diversity, fishing and hunting; for recreation;

M. Akmammedov (✉)
Minister of Water Economy of Turkmenistan, Deputy Chairman of the Cabinet of Ministers of Turkmenistan, 142 Hero of Turkmenistan A. Niyazov Str.,
Ashgabat, 744000, Turkmenistan
e-mail: khanmedov@mail.ru

C. Madramootoo and V.A. Dukhovny (eds.), *Water and Food Security in Central Asia*, NATO Science for Peace and Security Series C: Environmental Security,
DOI 10.1007/978-90-481-9974-7_19,

Table 1 Water use in the Aral sea basin

Datum	Unit	1960	1970	1980	1990	2000	2006
Population	mln. people	14.1	20.0	26.8	33.6	41.5	44.96
Irrigated area	ths. ha	4,510	5,150	6,920	7,600	7,990	8,456
Total water intake	km^3/year	60.61	94.56	120.69	116.27	105.0	166.30
Water intake for irrigation	km^3/year	56.15	86.84	106.79	106.4	94.66	95.97
Specific water intake per 1 ha	m^3/ha	12,450	16,860	15,430	14,000	11,850	11,350

Table 2 Collection and drainage water in the Aral sea basin km^3 a year

Country	CDW	Industrial plant's waste water	Total return water formed	Water discharge and utilization: To rivers	To natural depressions	Reuse for irrigation
Kazakhstan	1.6	0.19	1.79	0.84	0.7	0.25
Kyrgyzstan	1.7	0.22	1.92	1.85	0	0.07
Tajikistan	4.05	0.55	4.60	4.25	0	0.35
Turkmenistan	3.8	0.25	4.05	0.91	3.1	0.04
Uzbekistan	18.4	1.69	20.09	8.92	7.07	4.1
Total in the Aral Sea basin	29.55	29	32.45	16.77	10.87	4.81
Incl. Syrdarya's basin	11.95	1.44	13.39	9.16	1.54	2.69
And Amudarya's basin	17.6	1.46	19.06	7.61	9.33	2.12

and for other economic purposes. However, pollution from the mineral fertilizers and pesticides used in agriculture considerably limits the possibility of reusing the collector and drainage water for economic purposes. Collection, transportation, and diversion of drainage water involve huge expenditures of material resources.

Under present-day conditions, the generation of large volumes of CDW is related to the low efficiency of watering methods and techniques being used as well as crop irrigation technologies. According to scientific recommendations, with high-tech irrigated agriculture, the volume of drainage flow formed must not exceed 10% of the water volume delivered to the irrigated fields. However, due to tremendous non-productive losses of irrigation water, the actual share of CDW in the flow volume delivered to irrigation is 20–25% (Table 3). Taking into account that not all the irrigated area in the region is provided with collector and drainage systems,

Table 3 Collection and drainage for irrigated lands

Country	Volume of water resource use for irrigation, mln. m^3	CDW formed due to irrigation, mln. m^3	CDW percentage of flow quantity for irrigation, %
Kazakhstan	7,959	1,600	20.1
Kyrgyzstan	3,100	1,700	54.8
Tajikistan	10,150	4,050	39.9
Turkmenistan	16,788	3,800	22.6
Uzbekistan	56,660	18,400	32.5
Total in the Aral Sea basin	94,657	29,550	31.2
Incl. Syrdarya's basin	35,089	11,950	34.1
And Amudarya's basin	59,563	17,600	29.5

especially mountain districts where mainly the so called "dry drainage" operates, huge water consumption per irrigated hectare and substantial quantities of drainage flow are evident. This simply means that impractical use of the river water and transformation from good quality water to man-made polluted water is occurring.

Consequently, when elaborating measures for steady management of drainage flows, our efforts must be aimed first of all at water saving and universal reduction of nonproductive irrigation water losses and, accordingly, at reducing specific drainage water from an irrigated hectare and at lowering the total CDW volume.

In the initial period of intensive extension of irrigated areas and large-scale construction of collector and drainage systems, drainage water diversion to stem streams was widely applied so that water could be reused. Such "drawn and repeated" use of water resources specified in previous "schemes" and complex water resource use projects was justified by the need to increase the irrigation capacity of the river flow (available water resources). In the schemes of complex use of the Aral Sea basin's water resources due to repeated use of return water, drawn by design institutes, the forecasted possible increase of irrigation capacity of the river flow was 15–20%.

However, the development of irrigated agriculture in Central Asia in recent decades has shown that the "drawn and repeated" use of available water resources through stem streams is admissible only to a certain CDW return limit. Beyond this limit it may cause great damage to not only the drinking water supply, but to other sectors of the economy and particularly to the development of the agribusiness industry, and results in the deterioration of river water quality. In the upper reach, water salinity has increased by 0.2–0.3 g/L; in the middle reach, by 0.5–0.7 g/L; and in the lower reach, by 1.0–1.5 g/L. Rising water salinity leads to a decrease in crop yield. An increase in salinity of 0.1 g/L compared to the initial value causes damage to productivity estimated from 134 to 147 US dollars per hectare at the middle and lower reaches of the Amudarya River basin; and at the middle reach of the Syrdarya River basin the detriment amounts to 70–150 US dollars per hectare.

The influence of drainage discharge for the drawn-and-repeated use of water resources on river water quality can be illustrated by the example of the Amudarya

Table 4 Salinity at river stations along the Amudarya river (g/L) (V. Dukhovny)

Period	Atamurat (Kerki)	Ilchik	Birata (Darghanata)
1960–1965	0.56	0.62	
1966–1970	0.56	0.61	
1971–1975	0.67	0.7	
1976–1980	0.73	0.73	0.88
1981–1985	0.79	0.91	1.15
1986–1990	0.73		1.05

Table 5 Flow rate and salinity (mg/L) on the middle and lower reach of the Amudarya river

Date of water sampling	Place of water sampling	Average daily flow rate of the river and drainage collectors at the places of water sampling, m/s	Total salinity of solids, mg/L
02.06	Village Mukry, 1,102nd km of the Amudarya from the Aral Sea	2,300	600
03.06	Atamurat city, 1,045th km	1,880	633
04.04	Area of the head feeling of the Karabekaul canal, 950th km	1,840	633
04.06	Site Yapach, 5 km upstream from the South drainage collector inflow place from the Uzbekistan territory, 865th km	1,835	666
04.06	Mouth of the south collector, 860th km	30.0	5,700
04.06	10 km downstream from south collector inflow place, 855th km	1,860	933
07.06	Turkmenabad city, 840th km	1,850	733
07.06	Seidi city	1,840	736
07.06	Mouth of the Head left-bank collector, 724th km	23.4	1,400
07.06	Mouth of the Farab free-flowing collector, 746th km	3.0	1,766
07.06	Village Kabakly, downstream from the Makhankul collector inflow place from the Uzbekistan territory, 665th km	1,835	1,433
07.06	Village Birata, 611th km	1,830	800

River, where from 1960 to 1990 construction of collector and drainage systems and drainage water diversion to stem streams was widely developed (Table 4). The subsequent increase in water salinity along the river indicates the increase of pollution.

Analysis of observations for the 1986–1990 period compared to that of 1960–1965 shows that due to increasing volume of CDW discharge to the

Amudarya River basin, salinity at the Atamurat station had risen by 170 mg/L, at the Ilchik station by 290 mg/L, and at the Birata gauging station the salinity came to 1,200–1,300 mg/L.

The adverse influence of drainage discharge on Amudarya River water quality at the river's middle and lower reaches is also confirmed by the results of our monitoring carried out in June 2004 (Table 5).

According to the results of our observations in early June, water salinity at the Mukry village station (at the head intake of the Kurakum River) came to 600 mg/L. At the station 10 km downstream of the inflow of the South collector, river water salinity amounted to 930 mg/L, and that of the South collector was 5,700 mg/L. In the vicinity of the Kabakly village, downstream of the Makhankul collector discharge, river water salinity came to 1,400 mg/L. Apparently our samples were taken before complete mixing of the drainage and river waters, since at the site of Birata gauging station the river water salinity amounted to 800 mg/L.

As is apparent from this data, water salinity from the Mukry station to the Birata station increased by 200 mg/L at a water flow rate of about 2,000 m^3/s, which is a substantial value. After the discharge of drainage water, especially of the South and Makhankul collectors that have high drainage water rates with high mineralization levels, the water quality of the Amudarya River deteriorates considerably. This situation substantiates that drawn-and-repeated use of available water resources through the stem streams at existing volumes is unacceptable. It is therefore necessary to search for new collector and drainage water management and utilization approaches that ensure their efficient use without significant detriment to other water users and to the environment.

One effective way to reduce drainage flow and lower the cost of drainage water collection and transport is by using groundwater with low levels of minerals at the place where it originates or by moistening the root-inhabited soil horizon of agricultural crops by means of reverse regulation of the water regime of the soil and sub-irrigation. According to research data, at the location of low-mineralized (1–4 mg/L) groundwater at a depth of 1 m, it is possible to meet the water requirement of cotton by 30–70%; at the 2 m depth, by 15–30%; and at 3 m, by 5–20%. Thus, by artificial control of groundwater levels at depths of 1.2–1.5 m, it is possible to save significant amounts of irrigation water and lower the cost of drainage water diversion. Artificial control of groundwater level is achieved by locking drainage collectors at the open collector using control structures or in the closed network by means of shut-off-and-regulating devices.

Today, in addition to performing managerial and technical procedures to lower water consumption in irrigated agriculture and reduce the formation of drain and waste water as much as possible, the following options can be offered to control and place the drainage and waste water formed in the Aral Sea basin:

- Use of CDW at the sites of their formation for the irrigation of salt-resistant crops, including mixing those with river water
- Use of CDW in the area of main diverting tracts for the irrigation of salt-resistant crops on light sandy-desert soils and other soils

- Efficient export of drainage and waste water from cultivated areas to artificial accumulators, and use of those water bodies for fish breeding, recreation and sports purposes. Also, the diversion of drainage waste water to lakes in the Amudarya and Syrdarya River deltas can be provided for in order to make watered zones and conserve biological productivity and ecological balance
- In future, the option of desalinating drainage and waste water for different purposes, including crop irrigation

The experience of low-water years shows that the use of low-mineralized drainage water mixed with river water for crop irrigation in certain cases does not cause a decrease in crop productivity. The problem of using low-mineralized water for the irrigation of salt-resistant crops on light-texture soils has been thoroughly studied and there are a lot of scientifically substantiated recommendations. Therefore, involvement of collector and drainage waters in a unified system of water supply to agricultural crops not only helps to meet the water resources deficit but is also a reliable way to reuse return water repeatedly and eliminate its adverse effects on the environment.

As mentioned above, one of the typical ways to export and place collector and drainage waters is through efficient diversion and accumulation in topographic lows. It should be noted that small accumulators – drainage water lakes in small-scale local collector and drainage systems – usually have a shallow water depth and relatively sizeable water surface. Therefore intensive water evaporation takes place and as a result, the water in these water bodies becomes brackish, aquatic and desert vegetation becomes degraded, and the biological productivity falls. Sometimes in depressions with abundant wild vegetation, alkaline land occurs, barren and with a brackish lake in the middle of the depression. Consequently, when considering the option of placing collector and drainage flow in natural depressions, it is necessary to decide the optimum size of this accumulator, although the biological productivity of these man-made lake-accumulators is going to decline from year to year as salinity increases.

The project of the Turkmen lake "The Golden Age" in the Karakum desert can be cited as an example of the large-scale collection and placing of great CDW volumes. Currently, the total volume of CDW formed in Turkmenistan's irrigated lands is estimated to be up to 6 km^3, which is 20–25% of the water intake from sources to meet irrigation requirements and is in line with the up-to-date level of the agricultural industry. Allowing for the CDW arriving from the territory of Uzbekistan, the total amount of the CDW being carried through the territory of Turkmenistan comes to 11 km^3. To date, autonomous systems of CDW diversion outside the cultivated area have taken shape in the velayats (regions) of Turkmenistan. A part of the CDW was discharged to the Amudarya and Murgab Rivers but the majority was exported to inter-barchan depressions of the Karakum desert, flooding desert pastures on its way, deranging wells and water-accumulating takyrs.

A particularly tight situation emerged in the Dashoguz velayat, where over 65% of the annual CDW flow carried en route through interstate drainage collectors systems to the Sarykamysh Lake, forms in the territory of the neighboring velayats of Uzbekistan. During leaching and preplanting periods of the year, an abrupt increase in the Ozerniy and Darjyalyk drainage collector CDW flow quantity takes place, which exceeds design parameters by 1.2–2 times.

Passing above-design flow quantities, especially downstream of the Ozerniy and Darjyalyk collectors, results in intensive erosion of the riverbed; breaks existing bridges, gas pipelines, power transmission and communication lines and discharge facilities; floods an existing drainage system; aggravates the reclaimed state of irrigated lands and pastures, and causes considerable environmental and economic damage.

To eliminate the causes of the above-mentioned degradation processes and to fulfill Turkmenistan's commitments to stop the discharge of mineralized drainage water to the Amudarya River from Turkmenistan territory, it was decided to make the Turkmenistan Golden Age Lake. As a natural reservoir for the Turkmenistan lake, they have chosen the depression Karashor. Its extent is around 100 km; width 15–20 km; and its total capacity is estimated to be 132 km^3. The Turkmen lake is to receive drainage water from two main drainage tract systems, the Dashoguz inlet (north) and the Trans-Turkmen collector (south).

The south system is to export the drainage water from irrigated lands of the Dashoguz velayat and drainage water coming from the territory of Uzbekistan via the Ozernaya and Darjyalyk collectors. The south system is to completely export the drainage water from the irrigated lands of the Lebab, Mary, Akhal, and Balkan velayats. The system will export the drainage water of the Right bank and Left bank of the Amudarya's middle stream; will take the drainage water from the Djar, Main Murgab, and Central Tedjen collectors, as well as collector and drainage waters of the districts of the Kopetdag region.

The maximum flow of the Dashoguz inlet is composed of two flows, exported from the Daryalyk (60 m^3/s) and Ozerniy (150 m^3/s) with a total of 210 m^3/s. The Trans-Turkmen collector will cross the Turkmenistan territory in the submeridional direction from Deynau in the east to the Karashor depression in the west. The overall length of this collector is 720 km; the discharge in the head water is 123 m^3/s, of which 58 m^3/s is the CDW discharge of the right bank of the Amudarya River within the Lebap velayat. Maximum design discharge comes to 240 m^3/s.

The implementation of this project will allow a number of most important economic, ecological, and social issues to be managed and will transfigure the desert. In particular, a lot of local dead lakes with brackish water, formed owing to irregular drainage water discharges to topographic lows in the desert, will disappear forever. Orderly collector and drainage water diversion from irrigated lands will be improved, thereby the problems related to land swamping and salinization will be solved. At the areas of main collectors and main inlets, (total length over 1,500 km), watered zones will form where desert-based tree, shrub and herbaceous vegetation will rapidly grow and as a consequence, the carrying capacity of pastures will substantially increase. This will also enable the cultivation of salt-resistant crops in the collector and inlet areas to meet the needs of the local population.

Over the last 30–35 years, research institutions of Turkmenistan have investigated various aspects of using CDW as an additional source of crop irrigation water (cotton, rice, maize, sorghum, millet, etc.). Use of CDW with salinity of 3–4 g/L produced crop yields lower than the yield obtained through irrigation by river water by just 5–10%. Furthermore, when using CDW (2–5 g/L) for the irrigation of natural vegetation (ilak, chopan-telpek, the camel's thorn, shor-chair, etc.) on pastures, the productivity of its dry biomass increased more than twice (from 31 to 63 metric centner per ha). In the main collectors and the Turkmen lake itself fisheries will be developed and conditions created for migratory birds to stay and have a rest.

In recent years, a series of promising collector-drainage and discharge water treatment methods have been examined. Of those, studying the ability of higher water plants to absorb organic matter and oil products from water; intercept suspensions; extract biogenic elements, heavy metals, phenols, pesticides, and radioactive substances is of interest. Brushwood, a higher water plant is increasingly used as a biofilter in sedimentation tanks, ponds, and channels. This artificially made planting is called "bioplateau". Common reed grass, narrow-leaved cat's-tail, great bulrush, water hyacinth, morass-weed, and so on are the most promising among the higher water plants. The best water treatment results have been achieved by planting reeds and cat's-tail patches transverse to the polluted water streamline.

Application of the bioplateau method for the purification of collector-drainage and discharge water is the most environmentally sound and economical in comparison to other methods. This is why this method is preferred when developing the feasibility study of the Turkmen lake project and a bioplateau system is provided for in most arms of the uniting collector. Such a positive aspect of the collector system establishment as well as organized drainage-discharge water diversion should be noted. Because of the natural susceptibility to flooding of the hydrological regime of the Murgab and Tedjen Rivers and small rivers of the north-west Kopetdag slope as well as lack of sufficient capacity of the regulating reservoirs, disastrous floods occasionally happen at these water sources, with disorderly discharge of flood water to the desert area. Organized diversion of this flood water, which has minimum salt content, through a collection system to the Turkmen lake will positively influence the water quality in the collectors and the lake itself, flora and fauna of the environment, and the biological productivity of the watercourses and lake. And above all, these discharge waters will not be wasted due to evaporation and filtration and will be directed to the lake for future reuse.

Turkmenistan is located at the center of the Eurasian continent, and owing to its geographical position and climatic characteristics, our country is significant for migrating birds and is on the flyway which stretches from the arctic coasts of Western and Central Siberia to Iran, Afghanistan, India, and Africa. The nesting and wintering places of many types of birds occur within the bounds of Turkmenistan. Construction of the Turkmen lake, a major water structure with an area of 3,460 km^2, in the north-west part of the Karakum desert with a main drainage collector that is 720 km long, will create new, favourable environmental and feeding conditions for migrant birds which currently have a limited water supply.

Filling of the old Zapadniy Uzboy riverbed with water and crossing the Central Karakum and Western Karakum sandy deserts with a new collector channel will serve as a good landmark for migrant birds and promote the appearance of new nesting and wintering places of such water birds and waders like geese, common coots, cormorants, sandpipers, etc. Supplying the Central part of the Karakum desert with water will also favour some hoofed mammals, in particular, goitered gazelle, Ustyurt wild sheep, wild boar, otter, coypu, muskrat, etc., as well as promote the preservation and enrichment of the biodiversity of our country.

Creation of the Turkmen lake is very urgent for Turkmenistan. It will not only improve ecological conditions in the country as a whole and in the Turkmenistan part of the Aral Sea region, but will also increase water supply availability and feed conversion ratio of the pastures, raise crop capacity, and provide additional facilities to develop fisheries.

Hydraulic Structure Safety in the Republic of Uzbekistan

Timur Kamalovich Kamalov

Large waterworks facilities with a great number of channel and off-channel basins for multipurpose use have been built in the Republic of Uzbekistan. There are more than 90 waterworks facilities, more than 100 main canals of king-size length and 45 hydropower stations (HPS). The waterworks facilities have very important functions which affect economic, ecological and social spheres. The problem of their safe and reliable operation is therefore very topical. Furthermore the sustainable development of the country and safety of the population in the area under their influence depends on the reliability of such structures.

Today's main concerns are a result of the age of the structures (30–40 years), causing changes to engineering data related to sedimentation, seismic phenomena, landslides, torrents, snow slides, subduction of surging glaciers and other geodynamics impacts, and having insufficient emergency warning systems on the large waterworks facilities. Social and economic consequences and damage to property caused by destruction and damage of the structures are more than significant, and the possibility of breakdowns rise with prolongation of their operational life.

What We Can Tell About Conditions of the Waterworks Structures at This Moment?

Analysis of survey data showed that for recent years there is a dangerous tendency of increasing defects and damage that require urgent action. The main defects are concrete deterioration in zones with variable level, destruction of slope protection, drainage system clogging, and the retrogression of the river downstream For instance, there is destruction of concrete mounts of the head water and reinforced concrete constructions on numerous reservoirs (Charvakskoe, Djizakskoe, Zaaminskoe,

T.K. Kamalov (✉)
State Committee for Safety of Large-scale Hydro-structures, 7 Navoiy Str.,
Tashkent, 700011, Republic of Uzbekistan
e-mail: v-nadzor@sks.uz

C. Madramootoo et al. (eds.), *Water and Food Security in Central Asia*, 207
NATO Science for Peace and Security Series C: Environmental Security,
DOI 10.1007/978-90-481-9974-7_20,

Gissarakskoe, Karaultepinskoe, Kamashinskoe, Yuzhno-Surkhanskoe and others); and defects of antifiltration facilities on some reservoirs (Tuyamuyunskoe, Zaaminskoe, Dekhkanabadskoe, Kalkaminskoe and others); and retrogression of river in the tail-water of some waterworks (Verkhne-Chirchiksky, Sherabadsky).

There is concern about the conditions of intake channels on the reservoirs (Langarskoe, Gissarskoe, Karabagskoe, Kyzylsuvskoe, Karasuyskoe, etc.) caused by sedimentation. The condition of mechanical equipment which has a shorter operational life than the structures, significantly affects the reliability and safety of waterworks, including the safety of flood discharge.

The most dangerous situation is the loss of metal bearing capacity of the gates, the defects and damage of inserted and locomotive parts, as well as failure and deterioration of load lifting mechanisms. The technical condition of the mechanical equipment is unsatisfactory on the following water facilities: reservoirs – Kuyumazarskoe, Tupolangskoe, Khodjamushkentskoe and Sarmychsaiskoe; pumping stations – Kuyumazar, Kiziltepa, Khamza I/II, Amuzang I/II and DGPS; canals – Eski-Tuyatartar and Parkent, etc.; waterworks - Kharkhursky, Sharabadsky and others. Reconstruction of the Yuzhno-Golodnostepsky canal head and its central branch and updating of mechanical equipment of waterworks facilities are being implemented very slowly.

It should be noted that waterworks facilities are insufficiently equipped with control and measuring instrumentation and it is not possible to obtain complete and reliable information to ensure the control of the structure's technical condition. According to the project, 1,909 piezometers are foreseen on 25 large reservoirs from which 520 (or 27%) are not operational and 269 (or 14%) need to be repaired.

There are also problems of rock-dammed lakes among which Sarez Lake is the largest with a volume of more than 16 km^3. Other rock dams in the Shakhimardan River basin caused human toll in Uzbekistan and Kyrgyzstan in 1998 and we need to pay serious attention to the natural water bodies in the region. The control and supervision of technical conditions and safety of waterworks operation as well as urgent measures on their reconstruction to prolong their life are under permanent control by our state.

This determines the need to develop the normative legal regulation of issues related to safety of the large reservoirs and enforcing the law of the Republic of Uzbekistan "About safety of waterworks facility" in 1999. The main aim of the law "About safety of waterworks facility" is to ensure protection of life, health and property of citizens and enterprises, to prevent destruction of buildings and structures, erosion, dangerous groundwater level change and other damage caused by a waterworks emergency. Therefore this law covers all waterworks where destruction could result in an emergency situation that threatens lives and health, violates the work environment and vital activity.

The system of ensuring waterworks safety is based on international and national experience including the long time know-how of the country, is implemented under the law "About safety of waterworks facility" and it consists of the following:

- Separation of activity on ensuring the safety of waterworks operation between Government, executive government bodies, local governmental authorities and operational organizations

- Determination of main duties of organizations used waterworks
- Implementation of the state control on waterworks safety when they are being designed, constructed and operated
- Avowal of waterworks safety
- Conducting the waterworks cadastre
- Determination of responsibility for infringement of the law on waterworks safety

The main goal of government regulation on ensuring the waterworks safety is to prevent emergency by means of the followings:

(a) Improving the rules and regulations of design, construction and safe operation of waterworks
(b) Establishing public supervision for implementing the rules and regulations on waterworks operation by organizations using waterworks, and also for activities of other persons which could reduce the waterworks safety level
(c) Observing and analyzing conditions of waterworks and foundations in order to discover and repair defects, which could otherwise lead to full or partial loss of the workable conditions of the structure and to an emergency
(d) Personnel training for implementation of anti-damage measures and actions in the conditions of localizing and eliminating damages, and also supporting the needed material and financial reserves

The general principle underlying the law is the determination of the responsibility principle of the waterworks operating organization for ensuring the structure's safety and maintaining it in a safe condition. Responsibility for ensuring security by means of adopting the waterworks safety rules, by which the operating organization has to be governed, and for monitoring the operating organizational activity in this sphere is laid on the government regulation body. This means that the binding provisions of the normative legal base on ensuring dam safety are aimed at solving the main problems.

In recent years, with the introduction of government controls on waterworks safety and of the law of the Republic of Uzbekistan "About waterworks safety", more and more attention has been given to safety considerations when waterworks are being designed, constructed and operated, and this has resulted in major accidents being avoided.

The public supervision implementation of waterworks safety is laid on the state inspection organization "Gosvodkhoznadzor" by the Decree of the Cabinet of Ministry of 30 March 1999 No 143. By the adopted decree the following are the responsibility of the inspection body: the control and supervision of the technical state and safe operating of waterworks in consideration with their classification; declaration of waterworks safety; forming and conducting the waterworks cadastre; inspection of state and conditions of waterworks operation; development and approval of the code of practice on safety operating and criteria.

For this period the preparation of the normative law and normative methodical documents regulated waterworks safety ensuring, in consideration of their classification, functional peculiarities, design, operating conditions and special safety requirements. The supervision body is established and the work on safety declaring and the waterworks cadastre conducting is initiated.

The task-oriented surveys of the technical state of large and very important water management bodies by Gosvodtechnadzor were given special attention. For the latest period more than 233 were investigated, and the technical diagnostics of 43 bodies' condition was implemented. Observations of dam deformation on 42 reservoirs and waterworks by the geodesic methods are being implemented. Also other works relating to determination and elimination of damage risk factors affecting the technical conditions reliability and safe operation of large and especially important water management bodies have been implemented. The cadastre of 223 large and especially important water management bodies from 273 controlled bodies is prepared.

The permanent communication of operating organizations with regional authority on civil defense and emergency situations and the state control body is of great importance. The communication between these organizations about any dangerous changes in waterworks conditions and about threats of accident, to ensure the local warning systems is in constant readiness, and to prepare an anti-damage measures plan, is set.

It should be noted that ensuring waterworks safety depends firstly on staff skills in operating organizations, in construction and design organizations, the governmental control bodies, and also implementation of legal, normative law and normative technical documentations by them. Development and implementation of these documents, the personnel training in the enterprises and organizations and the state control are the most important task in the first stage of law realization. Implementation in our country of the governmental control on ensuring the waterworks safety, of diagnostics of waterworks technical conditions with their safety assessment, monitoring of large and especially important structures' safety and other activity ensuring their safe operation has reduced the risk level of their damage.

It is necessary to bear in mind that well-timed prevention of natural and manmade disasters is less costly and more effective than elimination of consequences from natural, technological and economical catastrophes. Any structure is subjected to impacts of natural, technological and temporal factors, which decrease its operational capability and reliability. It should be remembered that even the best irrigation and drainage system designed on the basis of modern scientific and technological achievements, and the most rational scheme will be less effective if the system is not maintained at a high technical level. Only the proper and rational maintenance allows the full use of the project and ensures the highest safety and reliability of all its elements during its operational life.

MIX
Papier aus verantwortungsvollen Quellen
Paper from responsible sources
FSC® C105338

If you have any concerns about our products,
you can contact us on
ProductSafety@springernature.com

In case Publisher is established outside the EU,
the EU authorized representative is:
Springer Nature Customer Service Center GmbH
Europaplatz 3, 69115 Heidelberg, Germany

Printed by Libri Plureos GmbH
in Hamburg, Germany